Library of
Davidson College

TO SERVE
WITH HONOR

ALSO BY RICHARD GABRIEL

The New Red Legions: An Attitudinal Portrait of the Soviet Soldier
The New Red Legions: A Survey Data Sourcebook
Crisis in Command: Mismanagement in the Army
Managers and Gladiators: Directions of Change in the Army
Ethnic Groups in America: The Irish and Italians
Program Evaluation: A Social Science Approach
The Environment: Critical Factors in Strategy Development
The Ethnic Factor in the Urban Polity

TO SERVE WITH HONOR

A Treatise on Military Ethics and the Way of the Soldier

Richard A. Gabriel

Foreword by
Vice-Admiral James Bond Stockdale, USN (ret.)

Greenwood Press
Westport, Connecticut · London, England

Library of Congress Cataloging in Publication Data

Gabriel, Richard A.
To serve with honor.

   Bibliography: p.
   Includes Index.
   1. Military ethics.   2. Soldiers—Conduct of life.
I. Title.
U22.G25   174′.9355   81-6254
ISBN 0-313-22545-1 (lib. bdg.)   AACR2

Copyright © 1982 by Richard A. Gabriel

All rights reserved. No portion of this book may be reproduced, by any process or technique, without the express written consent of the publisher.

Library of Congress Catalog Card Number: 81-6254
ISBN: 0-313-22545-1

First Published in 1982

Greenwood Press
A division of Congressional Information Service, Inc.
88 Post Road West, Westport, Connecticut 06881

Printed in the United States of America

10 9 8 7 6 5 4 3 2 1

Dedicated to the men of the U.S. Army who have served with honor, suffered the sting of battle, and died in the service of the republic. To Louis John Grieco, United States Navy, killed in action on May 11, 1945, at the age of twenty-one; and to my father, Alfred John Gabriel, USA, wounded in action outside Metz in November 1944 whose actions on and off the battlefield served as the inspiration for this book.

The question is this: do men, as a matter of ascertainable fact, want serious demands made upon their courage, loyalty, generosity and understanding? Do men, in other words, care to be moral beings, and do we prefer a life that might penalize us somehow for being craven, faithless and ungenerous?

—Walter Karp

We should be willing to asume that most men have sufficient desire to live a moral life, that they will profit from instruction that helps them to become more alert to ethical issues, and to apply their moral values more carefully and rigorously to the ethical dilemmas they encounter in their professional lives.

—Derek C. Bok

And is there anything more important than that the work of the soldier should be well done?

—Plato

CONTENTS

Foreword	xiii
by *Vice Admiral James Bond Stockdale*	
Acknowledgments	xvii
1. *The Ethical Climate*	3
Vietnam	3
The Academies	5
The All-Volunteer Force	5
The Search for Ethics	7
Issues	8
External Factors	10
Institutional Forces	11
The Cultural Milieu	15
The Challenge	19
2. *The Nature of Military Ethics*	23
An Ethical Perspective	23
The Nature of Ethics	26
A Definition of Ethics	28
Obligations	30
Sources of Obligations	33
Precepts	36
Art	38
Responsibility	40
Objections	42
Situational Ethics	50
Conclusions	53

x CONTENTS

3. *The Need for Military Ethics*   56
   - A Special Need   57
   - Military Effectiveness   58
   - The Vietnam Example   64
   - Negative Institutional Factors   65
   - After Vietnam and the AVF   68
   - The Organizational Malaise   70
   - A Cry for Help   73
   - Conclusions   77

4. *Professionalism and the Brotherhood of Arms*   80
   - The Role of the Soldier   80
   - Professionalism   82
   - The Military Is Different   86
   - A World Apart   88
   - Challenges to Professionalism   94
   - Responding to the Challenges   107
   - Avenues of Advance   111
   - Conclusions   116

5. *A Code of Military Ethics*   119
   - Advantages of a Code   119
   - Objections to a Code   122
   - Ethics and the Service Academies   131
   - The Ethical Dilemma   136
   - A Code of Military Ethics   138
   - Conclusions   147

6. *The Character of the Soldier*   150
   - Ethics and Virtue   150
   - Conclusions   173

7. *Loyalty, Obedience, and Dissent*   175
   - A False Fear   176
   - The Limits of Obligation   178
   - The Problem of Acquiescence   183
   - Legitimate Avenues of Military Protest   185
   - Obstacles   199
   - Conclusions   203

8. *Instilling Military Ethics*	206
Teaching Ethics	206
Institutional Forces	213
Enforcing Ethics	219
Conclusions	222
9. Final Thoughts	226
Bibliography	231
Index	239

FOREWORD

In this book, Richard A. Gabriel takes the lid off the potentially permanent damage the Vietnam era did to America's military might. As the ten-year war's political blunders, hardware drain, and even psychological impact on the man on the street fade into relative insignificance as part of the uniqueness of the event, major structural damage done to the bedrock of our soldiers' ethos remains. Two generations of American military officers brought up in the tradition of Robert Strange McNamara's business school ethic have unwittingly fostered a persistent and insidious warp in the age-old fighting man's value system, a warp that may forever inhibit victory on the battlefield.

In *To Serve with Honor*, Professor Gabriel describes the symptoms of this warp, explains how it came into being, and proposes a long-term cure. His thesis is an elaboration of what he stated in *Crisis in Command*, his earlier book written with Paul L. Savage, a book that set much of the civilian and uniformed military bureaucracy on its ear. According to that thesis, the entrepreneurial ethic and the military ethic are fundamentally opposed, and the latter has in recent years suffered badly from the invasion of the former.

Gabriel claims that the introduction of systems analysis, cost-effective criteria, and statistical measurement as overall tests of military value set in motion a process of erosion of the military ethic that seems almost impossible to check. The entrepreneurial system of values is inextricably tied up with successful management, company and personal profit, and rational self-interest. With these the military ethic is not at home. The soldier cannot adopt the methodology of business without adopting its language, its style, its tactics, and ultimately its ethics. Efficiency displaces honor as the greatest good. The military ethos is or should be one of duty, individual sacrifice, and group dedication. The traditional virtues of the military calling are loyalty, obedience, and courage. There may be lessons to be learned from the quarterly profit sheet of IBM or Textron, but these lessons should supplement, not submerge, military standards and values. That managerial values have been permitted to take over the front seat in the running of the military was the claim of *Crisis in Command*. The claim is repeated in Gabriel's present book, *To Serve with Honor*, augmented with a large-scale diagnosis of associated ills that have taken root in the value-structure of today's American society.

With regard to his central claim of military-managerial value conflict, I am much of Gabriel's mind. From first-hand experience in a war that I studied from both flak-encircled cockpit and bomb-encircled Hanoi, I can testify that the managerial spell wrought on the military bureaucracy has done injury that will be hard to repair. Our business school-oriented elite tried to manipulate rather than fight the Vietnam War. It would be good news to report that by the end of that conflict the entrepreneurial bewitchment of the military mind was broken. But unfortunately the spell still holds.

During my tenure as president of the Naval War College, I gave a course in philosophy and "ultimate situation" literature called "Foundations of Moral Obligation." Together we read many good essays by authors ranging from Aristotle to Jacques Monod. But of all the readings none stimulated more discussion than Gabriel's piece, "The Nature of Military Ethics," an essay that served the author as a preliminary sketch for *To Serve with Honor*. My students were officers in midcareer; they ranged in rank from lieutenant commander to captain in the Navy, and from major to colonel in the Marines, Air Force, and Army. Most were Vietnam veterans. Their nearly unanimous vote was that entrepreneurial standards and the military ethos are, as Gabriel claims, fundamentally opposed, and that the American military establishment has been invaded, to its detriment, by the standards and values of an efficiency-oriented ethic.

To those who claim that Gabriel's argument is one-sided and that his view of management versus military is absolutist, I would agree. But I would add that this world moves forward on the shoulders of one-sided people. It is those who say "But on the other hand" who bring up the rear—useful support personnel perhaps, but they are not in the vanguard. We should note that Gabriel himself does not claim that the entrepreneurial and military ethics stand glaring at each other from polar extremes. Rather, he tells us that there is a range, a spectrum of values along which the military ethic stands as much on one side of the midpoint as the managerial value-system stands on the other.

The symbiotic relation in which military leadership and industrial management find themselves today has a long history. Some would argue that the relation began before World War II. Few would deny that American industrial production was a weighty factor in winning that war. But few also would deny that by the end of that conflict the American military leadership was already finding the embrace of the nation's industrial complex a little uncomfortably tight. In his presidential "Farewell Address," Dwight Eisenhower warned against the dangers that could emerge from a too tightly locked-in "military-industrial complex." But the really damaging quantum jump in the business-military relation from a moral point of view did not occur until 1961 when Defense Secretary Robert Strange McNamara and his "Whiz Kids" took over the Department of Defense and tried to run the Vietnam conflict as if it were the Ford Motor Company with a knock in the engine and an unfavorable balance sheet. Among other lamentable results, direct or indirect, of this managerial ascendancy, Gabriel points to the exponential growth of careerism in the military, the loss of confidence by troops in their leadership, and the loss of that absolutely necessary ingredient of combat success, unit integrity.

Damage continued after the Vietnam War wound down to its inglorious close. Careerism persisted, fitness and efficiency reports grew even more inflated, readiness data, under the pressure of a pervasive bureaucratic "zero defect" mentality, continued to be falsified, juniors were scandalized by the indecisiveness, weakness, and—in some cases—outright dishonesty of their superiors.

Gabriel does not go so far as to make a demonology out of the McNamara influence. He knows that other factors are at work to the detriment of the integrity of the military. He does not believe that the All-Volunteer Force concept helps matters. With the closing down of obligatory military service, the armed forces lost the strength of a cross-section of the nation's youth. Now they must make do with the least highly qualified segment of the nation's young people. They have to deal with illiteracy, drug abuse, alcoholism, as well as with an increasing rate of desertion and criminality. Thus, vital energies that should go to improving military standards are inhibited at the start.

The most controversial section of Gabriel's book lies in his proposed honor code for the military. Other professions have their codes of ethics spelled out, he argues; why should the military be without one? It is true that, even before the American Revolution, there existed a tacit set of standards and values to which our military has tried to adhere as moral as well as professional ideals. These standards came to include the famous triad of the United States Military Academy—Duty, Honor, Country. But Gabriel finds these too generic, too comprehensive; he wants them spelled out in more detail. It is also true that since 1957 the military has had a code of conduct. But the code, however useful, aims principally at the special, though crucial, problem of the fighting man's responsibilities as prisoner of war. Gabriel wants a more comprehensive ethical code for the entire military.

There will be many in the profession who will question the value of a military ethical code. Codes make morality external, they say, and ethics becomes a matter of conforming to standards set up by others, not by oneself. Codes make ethics legalistic, they say, and may turn personal morality into courtroom procedure, into defense of one's "rights" when violations occur. If the provisions of a military ethical code are too broad, they may become platitudes and turn into the objects of lip-service. If they are too specific, a large book will be needed to list them, and who has the time to read, let alone practice, the precepts set down in it? Moreover, military philosophers since Epictetus have warned us about a proliferation of oaths.

Others will quarrel with the substance of Gabriel's posed code of ethics for the military profession. They will ask if it is for officers or enlisted personnel, or both. The way it is worded, they will charge, seems to set up an ambiguity hard to deal with—for the language of some of the provisions seems to be directed toward the specific responsibilities of officers who lead rather than those who look to them for leadership.

I share these difficulties and others too, but applaud Gabriel's willingness to tackle the tough job of formulating an honor code as well as the larger task of full-scale exploration of the relations between the military and moral obligation. This is an important book, one that will be widely discussed pro and con. It is a book that bears the indelible traits of its dedicated author, a military man and scholar with whom I have had the privilege of spending many educational and enjoyable hours.

Vice-Admiral James Bond Stockdale
United States Navy (Retired)

ACKNOWLEDGMENTS

This book would not have been possible without considerable support from others. St. Anselm's College deserves my thanks for granting me a sabbatical leave and thus allowing me to devote most of my time to the book. The freedom from classroom duties was truly liberating. I am grateful to the Brookings Institution where I held the post of guest scholar while I was researching and writing, and especially to Marty Binkin who first suggested my appointment. The National Defense University was most helpful in providing me an opportunity to present my ideas. Its president, and my friend, Lieutenant General Robert Gard is owed special thanks for first suggesting the seminar in ethics, an idea which at the time had far more critics than supporters.

Thanks is also owed to *Army Magazine, Parameters, America, Social Science, The Hastings Center Report, Military Review,* and the *New York Times,* all of which carried parts of the manuscript as articles. Their willingness to publish my ideas on ethics exposed them to a wide audience and, as a result of reader reactions, fired my enthusiasm for a more complete treatment of the subject. In this regard, special thanks is owed to Jim Binder, editor of *Army,* and Lieutnant Colonel Ernie Webb of *Military Review* who had faith in the work when it was only in its conceptual stages.

While this book was being written, several institutions provided forums for me to present my ideas and to benefit from the honest and forthright reactions of their students and faculty. I learned much from these sessions, and a great debt of thanks is owed to the institutions that made their halls available to me. They include the United States Air Force Academy, the Canadian Forces Staff College, Royal Roads Military

College, the United States Army Command and General Staff College, the Canadian Forces Training College, the United States Military Academy at West Point, the Army War College, the Royal Military College of Canada, and the University of Victoria in British Columbia.

No book ever gets written without the aid and comfort of one's friends and colleagues. I should like to express my most sincere thanks to John Keegan of the Royal Military Academy, Sandhurst, William Hauser of the William S. Paley Foundation, Colonel William Taylor of the United States Military Academy at West Point, Major Charles Cotton of the Royal Military College of Canada, and Colonel Dandridge M. Malone of the Army War College, all of whom read the manuscript and made suggestions. Whatever clarity of thought the book contains is largely their contribution. Special thanks is owed Major Steve Brodsky and Lieutenant Commander Don Lang, both of Royal Roads Military College, for their superior knowledge of military ethics in Canada. A significant debt is also owed to Louis Sorley whose work has had such great impact on my own. Paul L. Savage, my friend and coconspirator for some fourteen years, knows clearly the debt I owe him. Finally, my thanks to Admiral James Stockdale and Lieutenant General J. J. Paradis for providing examples of truly ethical soldiers. It is they whom I had in mind in writing much of this work.

Little could have been accomplished without the help of Terry Boissonneault who transcribed the manuscript, Anne Kugielsky who edited it, and Jim Sabin of Greenwood Press who published it. Lastly, but by no means least in my thoughts, are my wife Katherine and my daughters, Leah and Christine, who suffered through yet another book. I promise not to do it again... until the next time.

Manchester, New Hampshire
Winter 1981

TO SERVE
WITH HONOR

1 THE ETHICAL CLIMATE

Over the last two decades, the military has engaged in a good deal of soul-searching concerning the behavior of its members and of the profession itself. Those of us who served during this time are acutely aware of a deep sense of unease, a sense that the military may have lost its way. At the root of this sense of unease is the unspoken fear that the military may have lost its ethical compass. Many of the assumptions upon which military service rested, as well as many of the reasons for which military sacrifice was demanded, have become obscured.

It is almost a truism that we no longer know where we are going as a profession because we have not adequately understood where we have been.

Vietnam

One major reason for this sense of unease and lack of direction is clear enough: the Vietnam War. Into that battle zone, the American military brought with it the best equipped force in the world, the most highly motivated soldiers, and some of the best trained officers that the country could muster. But during that conflict the military discovered that its performance was seriously flawed, and it experienced a breakdown not only in its professionalism, but also in its ethical content. During the war, for example, drug use among American soldiers was epidemic. As many as 28 percent of American troops were chronic users of hard drugs, with as many as 600,000 soldiers becoming addicted during their tours of service.[1] This situation could hardly have escaped the attention of the officers in combat and staff positions; still the drug problem persisted.

Within combat units desertions and combat refusals were common. Both official and unofficial military accounts reported numerous incidents of units and men refusing to engage the enemy upon direct order.[2] Perhaps no one fact so clearly demonstrates the breakdown in military professionalism as the degree to which leadership elements, officers and noncommissioned officers (NCO) alike, were actually assassinated by their own men. As many as 1,016 officers and NCOs may have been killed by their own men.[3]

The insane body count doctrine added to the military's growing disorientation during the war. With the effectiveness of units determined by the number of enemy killed, officers had to risk the lives of their men to count the number of enemy dead following an engagement. As one general officer put it, "I shudder to think how many of our soldiers were killed on a body-counting mission—what a waste." Still another called it "A great crime and cancer in the Army in the eyes of the young officers...."[4] All the services began to resort to false reporting of performance statistics in an effort to please superiors. Thus, the American military began to perform the tasks of trivial bureaucratic housekeeping as a substitute for military effectiveness.

The experience of Vietnam created three crises for the military, which it has yet to overcome.[5] First, it suffered a *crisis of confidence* born of the war itself and fueled by charges of incompetence, cowardice, and mismanagement within the officer and leadership corps of the enlisted ranks. Serious doubts about the role of ground forces surfaced during that war and still remain. Second, the military began to undergo a *crisis of adaptation*. This traditional military organization is trying to come to grips with the revolution in America's societal values and life-styles that exploded in the 1960s, changes that have been deeply reflected within the military as well. Nowhere is this more clear than in the adoption of the All-Volunteer Force (AVF) concept. Finally, the Vietnam War produced a *crisis of conscience* stemming directly from a range of individual and organizational behavior during the war that has profoundly tried the conscience of the profession itself: charges of war crimes at My Lai; a series of coverups involving high-ranking officers, NCOs, and even generals; and a series of post exchange (PX) scandals. Excessive careerism practically became a hallmark of the officer corps during the war and remains so today. What is more, many cases of official misconduct were uncovered in which some officers were charged and found guilty of misusing their position of special trust.

These crises brutally demonstrated to many concerned members of the armed forces that the profession itself had changed drastically, and by and large, not for the better. Many officers now fear that the certainties that underpinned traditional military values are being eroded, and that the replacement values are less than satisfactory. There is a feeling that something has gone seriously awry and that traditional values have been replaced. The new set of values seems to have only tangential relevance to the actual workings of a military force on the battlefield. Hence, many of the professional soldiers and conscripts of the last decade no longer know the direction they have to travel in order to remain true and faithful to the obligations they undertook when they first entered service.

The Academies

Compounding the military's problems was the series of scandals in the 1960s and 1970s in the military academies in which cadets were dismissed for cheating, stealing examinations, and blatantly violating the honor code. The question that arose was whether ethics was being taught at all in the academies. Clearly, the number of classroom hours devoted to matters even remotely related to the question of ethics was insufficient. The cheating and honor code scandals and their attempted coverups seemed merely to drive home the point that the military profession had failed to instill a basic sense of right and wrong in its future leaders.

This situation could hardly have been helped by the complicity of a West Point superintendent, Major General Samuel Koster, in the massacre at My Lai. In 1970 General Koster was found guilty of complicity in that incident and was required to resign from service. Suddenly, military men of both academy and nonacademy background could no longer regard the long cherished academies as their ethical models.

The All-Volunteer Force

The problems of the military neither began nor ended with Vietnam. In 1973, the entire military structure was shaken to the core by the adoption of the All-Volunteer Force. The military profession itself was asked to make a complete role-change. Although military service would still be regarded as a special calling, it would no longer be inextricably linked to the rights and privileges of citizenship. With the adoption of the AVF, the military was to be treated as any other occupation, competing with the civilian sector in the marketplace to attract adequate manpower and quality. While managerial tendencies had been evident in the military since World War II, the appointment of Robert McNamara as secretary of defense in 1960 represented a consistent effort to institutionalize within the military a wide range of organizational practices and values that had their origins in the modern business corporation. With the adoption of the AVF, this process had come full cycle.

It did not take the professional military long to realize that the AVF was a failure. The social composition of the All-Volunteer Force was so out of joint with the rest of American society that it placed an unfair burden of service, and ultimately combat death, upon the lower socioeconomic strata of society. Blacks, minorities, and especially the poor were overrepresented by a rate three and four times their percentage strengths in American society. Furthermore, many men joined the AFV because they had no other economic skills with which to bargain.[6] Finally, military effectiveness began to suffer seriously, as the quality of raw recruit material fell disastrously. For example, the reading level of the average soldier dropped from the twelfth-grade level in 1973 to the fifth-grade level in 1980, or about one grade level per year; in 1979 42 percent of the soldiers were inadequately trained for the military occupational specialties in which they served; and desertion rates were exceedingly high, and drug use was still endemic.[7] As a result, the ability of the AVF to execute its military mission became questionable. For example, over 50 percent of the soldiers in the AVF fall into Category 3B or below, a category that the military reserves for the truly marginal soldier. The forces of the marketplace have replaced patriotism and citizenship as the basis of military service. The conscription process by lot has given way to conscription by hunger and lack of opportunity, and the fighting ability of military units has fallen off drastically as a consequence.

For many members of the military, especially career professionals, military service is not merely another economic choice, nor is it totally an individual obligation. They recognize that military service is a special community obligation based on the principle that membership in a community, especially a democratic community, implies the willingness to defend it. This link between the rights of the individual as citizen and his obligation to defend the community has always been among the greatest strengths of democratic politics. Moreover, it implies a certain responsibility for the lives and welfare of others in the community. The AVF struck at the very heart of these principles.

Members of the military contend that the determination of who enters military service cannot be allowed to rest purely on economic calculations of supply and demand. To do so would undermine a fundamental value of democracy: that all men count equally before the community. As a social institution, the AVF imposes the burden of possible death in battle on an increasingly narrow segment of American society—its poor, its minorities, its ill-educated, and its cumulatively disadvantaged. It marks the return to the social worth doctrine of earlier times when the privileged members of the community could escape military service altogether or hire substitutes. Many military professionals believe that the AVF can only cause serious rifts in the social fabric. There is the strong sense that democratic values and habits cannot long endure in the larger social order when the ultimate risk of membership in that order, the risk of death on the battlefield in its defense, is not equally shared by all segments of the society.

The Search for Ethics

The military profession realizes that whatever sense of ethics and professionalism it has clung to over the preceding decades needs reexamination and clarification. This reexamination and clarification would constitute the first step in a moral renaissance aimed at rediscovering the moral bearings of the military profession. To make this assertion is only to point out what those who have served in the profession throughout history have always known: that the effectiveness and success of a military force rests far more on the moral quality of its officers and men than it does on technical expertise.[8] Any profession requires at its center a sense of ethical certainty. By 1980, the American military profession was deeply involved in a search to rediscover its professional moorings and ethics as a way of regaining the military professionalism that has marked our military throughout most of our history.

While the military may be unsure of its own professionalism and ethics, it has certainly been in the forefront of the efforts to examine and clarify them, as is evidenced by the number of articles on ethics that have appeared in various military publications over the last decade. In addition, a series of official studies such as the Army War College Study on Military Professionalism have been conducted. Project studies performed by officers at the upper level staff service colleges openly and frankly explore the problems of ethics as they relate to their respective services.

The number of young officers who have been willing to speak out, perhaps even at risk to their careers, in article after article over the last decade testifies to the growing desire of the officer corps to be in the vanguard of an ethical renaissance. Thus, while the military does indeed have its shortcomings, a refusal to examine and document its own failures and to attempt solutions to its ethical problems is not among them.

Unfortunately, however, its effort has been without focus. The problem is so complex and the approaches to it so scattered that we are more confused now than we were a decade ago. In one sense we are more confounded by having been forced to rid ourselves of our old values, having seen them attacked and compromised from within, while we have yet to clarify new ones. As such, the issues remain confused and the solutions more elusive than ever.

Issues

Two basic issues account for this confusion. The first involves a tendency to confuse the ethics of the profession with the ethics of the good moral life. Military commentators have tried to make their ethical precepts totalist in nature, developing a set of professional ethics which are meant to extend not only to the military man's professional role but also to his life outside the profession. This is a confusion of the first order and will be dealt with throughout this book. Part of this issue is the tendency to confuse the good military professional with the good man, as if by being a good member of a profession and living up to its ethics necessarily constitutes a good moral man. The evidence from history that this is false is clear enough. The medical practitioners who applied their arts in good professional fashion yet killed hundreds of children in the concentration camps may have been good professionals; they were not ethical men.

This issue has led to a second, which is at the heart of the military's difficulty in clarifying its own ethical precepts: the tendency to confuse an *ethics of virtue* with an *ethics of duty*, a tendency to confuse the presence of certain traits of character with ethics. The classic example is the argument that if a man has integrity, he will always behave ethically. The British and the Canadians are very partial to the argument that the proper subject of ethical analysis is character and its inculcation in the young officer.[9] More will be said about this matter later; suffice it to say here that the possession of desirable character traits in no way guarantees ethical conduct. More importantly, one needs a very clear statement of the ethical obligations that one ought to observe if one is to be expected to behave ethically.

These issues have led to a debate about military ethics within the profession itself. It is a debate that has created more heat than light, however. These are not the only issues raised by the discussion. Indeed, the confusion and debate over the basic issues have led to even more confusion in other areas. Inevitably, once the military opened a Pandora's box by examining its own standards of professional conduct, many of the old standards were brought into question. Furthermore, many of the old questions which were thought settled were raised anew.

Some of the more critical questions that need to be addressed are the following: Should the military have a code of ethics? Can military ethics be taught, or, indeed, are ethics merely absorbed from the larger society, and, therefore, when one joins the military one brings with him, an assortment of ethical baggage from which there is virtually no escape? If this is true, what are the implications for sustaining military professionalism and ethical behavior within the profession itself? Are Western civilian values fundamentally antithetical to those values required for an effective military profession? To what degree can a military profession be separate from its society without becoming a threat to it? Is there a set of values appropriate to military service that is inappropriate to civilian society and vice-versa? Finally, what are the limits of loyalty to the profession and to the state? Have we confused loyalty with blind obedience and then camouflaged this obedience and made it a virtue on the premise that an unquestioning military that executes all orders of civilian authorities simply because they are civilians is less of a threat to the democratic state? While the answers to these questions may be less than ideal, they can be, answered properly only when they are rendered in the context of a set of ethical precepts appropriate to the military profession. The search for moral bearings must first be directed at the establishment, definition, and clarification of those ethical precepts that orient the military in the exercise of its critical function for the democratic state.

The concern for ethics has been greater in the military than in other professions. Approximately ten times as many articles dealing with ethics have appeared in military journals in the last decade as have appeared in the journals of the legal profession. Eight times more have appeared than for the medical profession, and twelve times more than have appeared in educational journals.[10] This is not to say that other professions have not expressed concern over many of the same ethical problems as perceived by members of the military. For example, a study undertaken in 1976 by Harvard Business School[11] on Watergate found that 59 percent of the business managers questioned about the behavior of their subordinates agreed that "the junior members of Nixon's reelection committee who confessed that they went along with their bosses to show their loyalty is just what young managers would have done in business." Indeed, 63 percent of top industry managers agreed that this type of behavior could also be expected of their subordinates. Accordingly, the military's concern about the problem of subordinates failing to tell the truth out of fear of damaging their own careers is echoed in the business community. In the War College survey, the legal, medical, and eductional professions had by and large the same concerns as members of the officer corps.[12] Approximately 80 percent of the military officers and medical personnel reported that they should live up to an absolute standard. Almost 50 percent of the legal profession as well as the military agreed that a gap existed between the ideal and practical standards of the professions. Another 50 percent believed that ethical conduct was not well regarded by the organization and that both the medical and military professions were likely to punish ethical conduct more than reward it. Over 50 percent of the members in these professions felt that their superiors were more concerned with obtaining results than with how results were obtained. Thus, careerist pressures appear to be equally present in all the major professions. Finally, 43 percent felt that the pressures of "the system" forced resort to questionable practices as a matter of survival.

External Factors

A number of external forces have had great effect on how the military profession defines itself. One of these forces is the shift in standards of morality and permissiveness evident in the larger society and coupled with liberalization campaigns for greater individual rights and privileges within the military itself.[13] American society is entering the post-industrial age with its attendant emphasis upon liberties and rights. In addition, the adoption of the All-Volunteer Force has forced the military to deal with traditional problems of discipline and cohesion with new policies and techniques that are more appropriate to civilian business enterprises than to a military organization. This liberalization has created a series of severe challenges at all levels of command and leadership.

The second external force centers on the tendency toward the civilianization of the military. The values of the civilian way of life are replacing those of the military; moreover, both the general citizenry and political leaders now perceive the way of the soldier as just another occupation. Since the advent of the AVF, most enlisted soldiers have been encouraged to regard the military as simply a job deserving no more loyalty and sacrifice than one would expect to find in a business corporation. Hence, there is a denial of any kind of higher level of discipline and sacrifice historically attendant to military service. The third force has to do with the injection of managerial values and entrepreneurialism into the military. With the appointment of Robert McNamara as secretary of defense, the U.S. military became increasingly bureaucratized. The bureaucratic mechanisms he set up were much like those found in the business corporation, and they were grafted upon the military profession with little thought as to their effect. While these mechanisms were helpful in the development of systems supporting the combat functions, they were absorbed right down to the troop unit level, to the point where they are now being used to "manage" men rather than lead them.[14] These civilian values tend to erode the values and habits that are at the root of the military profession. For example, much of the lack of unit cohesion found among U.S. combat units in Vietnam may well have stemmed from the lack of leadership resulting from the institutionalization of a series of managerial practices. As a consequence, the interpersonal bonds between leader and led, which are so important to the combat effectiveness of the small unit, became eroded, and many officers were forced to choose between their roles as military professionals and managers. Too often, those who chose their professional roles as leaders were set back in their careers.

Institutional Forces

In addition to these external elements, certain institutional forces have given rise to major ethical challenges from within the profession. According to Kermit Johnson, there are four such institutional forces.[15] The first of these is a pervasive sense of ethical relativism, the value premise that if something works and gets results, then it is right. Nothing is more important than success, and nothing succeeds like success. Major General Koster, former commandant at West Point, clearly reflected this attitude when he advised his cadets that "we are interested in the doer, not the thinker." While the military has always embraced the notion that if some course of action works then it is right, the doctrine of ethical relativism perverts this traditional "can do" attitude. As a result, getting to the top has legitimized virtually all roads and means of getting there.

The second institutional force which has posed ethical challenges to the profession is the "loyalty syndrome," a condition in which subordinates are loyal to their superiors to the point of concealing their superior's shortcomings or never questioning their judgment. The "loyalty syndrome" raises the condition of loyalty to an imperative duty, so that the execution of even wrong or silly orders becomes perceived as part of the highest value of the profession. When this happens, loyalty becomes perverted and placed at the service of crass careerism.

A third factor is the excessive concern for image in order to protect the profession from embarrassing information. All too often (as during the Vietnam War) this leads to concealing mistakes, at times to the point of criminal behavior. The tendency is not to "rock the boat" and not to submit efficiency or readiness reports that might reflect less than perfect performance.

Finally, there is the "drive for success," which tends to manifest itself especially within officers who have been corrupted by the values of entrepreneurship, careerism, and managerialism.[16] It is not surprising that, as the military began to integrate the values and techniques of the business community, the inordinate drive for career success so characteristic of the "successful" business executive began to afflict some of its officers. Such an affliction is clearly evident today even to those who have served the shortest time.

Not all of the military's ethical crises can be blamed on business. The military's own behavior has destroyed many of the institutional supports for ethical action. Take, for example, the disastrous doctrine of "zero defects" which is familiar to most officers. The idea of zero defects was borrowed directly from civilian business and is related to mass-produced consumer items. It was transplanted directly to the military as a managerial device for reducing errors in performance. The difficulty with the zero defects doctrine is that it is impossible to implement in human behavior, for no human being is ever completely free of defects. Trying to enforce a policy of zero defects produces in the military a "zero defects mentality" in which one cannot admit any failure in trying to implement the policy itself.[17] One practical consequence has been the falsification of reports on anything from unit readiness to ammunition supplies on the grounds that it is worse to admit a mistake than to make one.

In other instances, the profession sends clear signals to the membership that ethical judgment is not to be trusted. For example, an officer is required to produce two sets of identification while in uniform to enter the post exchange or to produce a marriage license to qualify for his dependent's allowance. The only way to make an officer ethical is to give him the opportunity to act ethically, and this implies the opportunity to act unethically as well. To continue to suggest that an officer's word is his bond is nonsense when he is required to produce identification to cash a check at his own bank or to produce a birth certificate for his children in order to secure adequate housing. These are institutional conditions which attempt to substitute bureaucratic procedures for ethical judgment and responsibility. The existence and use of these institutional conditions lead to a reliance upon bureaucratic rules and mechanisms of control, while undercutting the soldier's opportunities to exercise ethical judgment.

Other institutional devices, such as centralizing promotions, reducing the power of unit commanders to enforce discipline, and encouraging rapid turnover in command positions, all reduce the opportunities for ethical judgment. Such institutional devices are felt most severely in the personnel evaluation reports. Every officer knows that a single less-than-perfect efficiency report may well mean the difference between a successful and unsuccessful career. The same is true for the noncommissioned officer. As a consequence, leadership elements too often stress the "wolf-at-the-door syndrome." In assuming their positions, they take few risks, try to keep the lid on existing problems, and avoid confronting problems directly. To admit that a unit has problems is often perceived as a reflection on the commander's abilities. Thus, commanders often hope to keep everything in order, at least publicly, long enough to receive a good efficiency report and then move on to another assignment. This manner of viewing command responsibilities often forces an individual into a series of moral confrontations with his own sense of ethics and brings that sense into direct conflict with the organizational imperatives for success.

All too frequently, the imperatives for career success and the desire to advance win out.

Of course, men in the military are no more or no less ethical in theory than those in other professions. It may be, however, that the task of the military profession is so different and the responsibility so awesome that the ethical standards that are acceptable to society at large, and even to other professions, are simply insufficient for the military professional. Other more exacting standards may be needed.

The U.S. military has historically lacked a specified code of military ethics, as well as specific instruction in the ethics of the profession. In dealing with ethical problems, the military officer, therefore, finds himself confronted with three major handicaps. First, he has little training in recognizing ethical dilemmas. Many times a soldier simply does not perceive problems in ethical terms, or even recognize obvious ethical dimensions associated with them. In contrast, the medical and legal professions spend considerable time training their novices in the ethical problems of their profession, while at the same time providing their practitioners with a specific code of ethical conduct.

Second, few military personnel have any training in moral reasoning. This is not surprising considering the lack of training in ethics and the lack of a formal code. When it is assumed that the ethics of the military are essentially the ethics of the larger society, and when it assumed that a soldier brings to the military a set of personal ethics that is adequate for military life, then it is not too difficult to understand why the military professional is given little training in moral reasoning by the profession itself. In those rare instances when a soldier does recognize moral dilemmas, he finds it difficult to develop morally acceptable solutions because he is not used to reasoning in moral terms.

These two handicaps produce a third: a tendency to resolve those few ethical dilemmas that are perceived in favor of organizational imperatives. When in doubt, the tendency is to adopt a "can do" attitude and to carry out orders even if one has serious ethical reservations about them. It is unrealistic, however, to expect men who have not been exposed to an ethical code and who have not been trained in moral reasoning to do anything else except resolve the few ethical dilemmas they do see in terms of the demands of the organization. That course of action at least gives the individual soldier the false security of being among many of his peers and, perhaps, even of being in a majority. Any other course of action predicated on moral grounds which runs contrary to the organization's norms may force him into the solitude of being in a moral minority, perhaps even a "majority of one," at war with the profession in which he claims special membership.

The Cultural Milieu

That the military has been buffeted by the winds of political, economic, social, and ethical change is clear enough. This dynamic is directly affecting the military not only from without by near-revolutionary forces, but also from within by the implementation of institutional changes. While the profession must share culpability, the fact is that the military is only one of many social institutions awash in the societal sea. Although some aspects of the profession's behavior can be—indeed, must be—successfully isolated from the values of the larger society, it must be understood that some societal forces may be hostile to its professional ambience. This is true in all societies to some extent, especially Western democratic societies, but particularly so in the United States where major cultural forces have actually impeded the development of a sense of ethics appropriate to the military profession.

The military's inability to develop a code of professional ethics is not very surprising in the context of American social values. The major values that have shaped so much of American society work against the development of *any* ethical codes that serve community as against individual interests. The magnitude of the difficulty is, therefore, apparent in developing professional military ethics. The basic values and operating assumptions of American society in its economic, social, and political development have made it all but impossible to develop notions of community service that transcend individual self-interest.[18] For example, American economic life is predicated on the assumptions of the free enterprise economics of Adam Smith. These assumptions posit, in effect, that the greed of each equates to the good of all. Accordingly, society is seen to be comprised of a number of individuals pursuing their economic self-interest. Through this pursuit of self-interest it is assumed, as if by an "invisible guiding hand" or the "forces of the market place," that some approximation of the common good will result. In economic terms, it is affirmed that there is no fundamental tension between the pursuit of "legitimate" self-interest and the common good.

Indeed, the common good is seen to result naturally from the pursuit of individual self-interest.

This same premise is at the base of the American social system, with its roots struck deeply in Social Darwinism. Here the assumption is that individuals who rise to the top of the societal pyramid do so because they are "the best." Drawing upon an incorrect biological analogy in the work of Charles Darwin, Herbert Spencer argued that the societal struggle rests in the survival of the fittest, with the weak, if not eliminated from the social milieu, at least relegated to the bottom of the social heap with its attendant costs. Social Darwinism contains a basic assumption of American social life, that every individual pressing his self-interest will in the process produce a social order that is relatively just. A society rooted in Social Darwinism places fundamental emphasis upon the role of the individual and the pursuit of his ends, and is antithetical to the development of communal interests in which individuals are expected to forego self-interest in the service of larger community goals or values.

A variation of both Adam Smith economics and Social Darwinism is found at the base of the American political system with its theoretical and institutional roots struck deeply in the political philosophy of James Madison. In Madisonian political thought, it is assumed that individuals will press their own interests no matter what government does to curtail this drive. The tendency to form groups in the pursuit of similar interests is, as Madison noted, sown into the very soul of man. The problem for Madisonian politics, therefore, is how to control the tendency of individuals to pursue self-interest to everyone else's potential detriment. There are only two real alternatives. One is to restrict liberty so as to forbid the pursuit of self-interest, and the other is to devise a governmental system that can control the effects of this relentless human drive by limiting their impact through institutional means. As a democrat, Madison could not choose to restrict liberty. Instead, he chose to structure the governmental process in a Byzantine manner. The structure of checks and balances and separation of powers would allow groups to press their self-interest, but the overly complicated process of government would modify individual demands, cancel others, and moderate the more extreme demands. The end product of the democratic process would be some semblance of the common good. However, there is never any question in Madisonian politics that the pursuit of individual self-interest constitutes the *raison d'etre* of the political order. There is only the assumption that a number of individuals pressing their interests could somehow be made to press them within the extremes of political action.

When taken together, it is clear that these three major intellectual and normative strains of American social development—the economics of Adam Smith, the social ethics of Herbert Spencer, and the politics of James Madison—are structured around the proposition that there is no fundamental tension between individuals pursuing their own subjectively defined notions of self-interest and the larger notion of the common good. Thus, it is not difficult to comprehend why professions like the military, themselves subsocietal communities, have found it so difficult to develop notions of community ethics wherein concerns of self-interest are appropriately and morally subordinated to the larger interests of the social and ethical community that constitutes the profession. The concept of a sense of ethics which is applicable to the community and which conflicts with individual self-interest is, in the face of the larger intellectual thrust of American society, seen as either unnecessary or impossible. An ethics of the community is regarded as unnecessary because whatever communal goals may exist are assumed to result naturally from the process of pursuing private interests. Accordingly, as long as individuals pursue their self-interest there can never be, by definition, a failure to serve community goals. The only other logical alternative is to affirm that the pursuit of community goals which require the sacrifice of self-interest or are antithetical to it is impossible since the only point of theoretical reference remains the individual and not the community. In this sense, communities do not pursue goals; they are only "convenient fictions." Only individuals in an empirical sense pursue goals, and, therefore, the only point of ethical reference is the individual's interests. In a society structured around these fundamental premises, it is not surprising that Americans are far more at home with notions of self-interest than with ideas of community, with rights than with duties, and with personal ethics than with professional ethics.

With regard to its application to the field of ethics, Smith's economics, Madisonian politics, and Spencerian social ethics lead directly to a fourth intellectual strain, ethical egoism. Ethical egoism is the philosophical view that affirms that an individual ought always to do that which will promote his own greatest good. Good is defined subjectively as that which promotes the individual's self-interest as he sees it; evil is seen as that which does not advance the individual's subjectively defined self-interest. When applied to the discipline of ethics, the pursuit of ethical egoism is the ethical equivalent of the assumptions that underpin Smith's economics, Madisonian politics, and Spencerian social ethics. Ethical egoism does not perceive a tension between individual interests and community interests since the latter is not really presumed to exist apart from the concrete interests of individuals.

The ethical egoist's most basic and only obligation, therefore, is to promote the greatest possible balance between what he sees as good for himself and what he defines as bad for himself. The point of reference is solely and totally the individual, and it clearly implies the negation of larger communal interest. The ethical egoist may hold any kind of theory as to what constitutes good or evil as long as he believes, or at least perceives, that it is in his interest to believe that something is good or evil. The point is that for the ethical egoist there is no clear standard *external* to the individual to which he can be obligated to comply.

Ethical egoism is firmly rooted in assumptions of psychological egoism which see man as by nature always seeking his own good over that of other men. Madisonian politics shares this same assumption, as do Spencerian social ethics and Smith's economics. Psychological egoism affirms that an individual can have concern for the well-being of others only as a means of furthering his own welfare, never for its own sake. Ethical egoism does not necessarily negate the proposition that there may be other communities to which individuals may owe allegiance, but it does affirm that the individual is under no obligation to support these communities unless it is in his own self-interest to do so. It is clear, therefore, that for the ethical egoist professions like the military which require individuals to observe obligations to the community *instead* of their own self-interest rather than in addition to it will appear illogical. Thus, ethical egoism as a logical corollary to other value strains in American culture works at cross-purposes to developing codes of ethics for any profession, especially in the military where death on the battlefield in defense of community interests may work strongly against perceived notions of self-interest!

Because Americans have placed the focus of the ethical equation so strongly upon notions of self-interest, it is not surprising that our military establishment is sometimes characterized by careerism and powered by economic analogies that military service is the equivalent of civilian occupations. Nor is it surprising that we have denied the existence of special obligations among members of the profession, or that we have seen a tension emerge within the military between the pull of vocation and occupation. Nor is it surprising that we find ourselves, especially since the adoption of the AVF, unable to socialize new recruits to the military way. Other analysts have pointed out some of these problems, but they have often attributed them to the wrong causes. These problems are often seen to stem from the organizational structure's failure to come to grips with the ethical perspectives of the individual when the real problem is exactly the reverse: the inability of the profession to develop a sense of community ethical obligations that would take precedence over those of the individual.

As long as the focus of military ethics remains tied to notions of individual self-interest and as long as ethics are viewed as only descriptive devices instead of prescriptive and proscriptive rules requiring observance, it will be impossible to develop a sense of community obligation for the profession. And the problem will become as circular as it is vicious. Without a clear statement of community obligations, the soldier cannot be socialized to a sense of community ethics, and without a sense of community ethics we cannot develop a sense of community obligation. The first step in breaking the circle is to recognize that the common assumption of Smith's economics, Darwinian social ethics, Madisonian politics, and ethical egoism, that the pursuit of self-interest will magically produce a sense of community values, is simply a false one. This false assumption eats at the very heart of the community obligations which are so necessary to any code of ethics.

The Challenge

The challenge of military ethics is to develop a specific code of professional ethical precepts that will take precedence over individual ethical orientations and to teach the code effectively. The beginnings of the code require the realization that the military profession is different from all other social institutions. The military's primary function, after all, is that of organized social violence in which it demands the sacrifice of the lives of its membership in pursuit of the community's right to self-defense. Until we realize that the military is a qualitatively different type of social institution we will be unable to evolve a code of ethics for it. Until we evolve such a code we will not be able to teach it, and until we teach such a code we will not be able to reenforce the idea among the membership that what they do is fundamentally different from other tasks in the social order. This is the central challenge of any treatise on military ethics, to evolve a specific code of ethical precepts for the military profession and to delineate the philosophical basis for it.

In teaching a code of ethics, a problem of the first magnitude is finding qualified instruction. In the past, as Derek Bok has pointed out, most teachers of ethics have had few qualifications beyond a strongly developed social conscience. Indeed, the quality of moral reasoning has little to do with how intensely one feels about the subject; the fact that one feels intensely may actually become an impediment to clear thinking. Moreover, teaching ethics must be done correctly and well. A badly taught ethics course will only confirm the persistent notion that moral reasoning is both impractical and inconclusive.[19] Teaching ethics in the military is even more of a problem when military staff colleges revise their curricula to focus on matters that are only tangentially related to the arts of war.[20] The lack of professional orientation in education, coupled with unfocused curricula, has made it very difficult for the military to identify a coherent core of subjects importantly related to ethics. This failing apparently characterizes the military academies as well.

The problem of developing an ethical code and teaching it is more difficult in the military than in other professions. In the first place, the military has never had a specifically defined code of ethics which has been clearly postulated and against which it has been willing to measure its own behavior. Moreover, the military has no tradition of producing from within a corps of ethicists who could teach ethics, because it has relied upon the society at large to instill ethics in its membership and has never felt it needed ethicists. Until recently, the profession has not seen the need for an ethical code either. As a result the profession has traditionally produced neither. Furthermore, the military generally regards those outside it with some distrust. Thus, it fears hiring academics to teach its courses on the grounds that they are likely to be unsympathetic, if not antimilitary, and it fears relying on the philosopher because it suspects he is impractical.

Compounding this problem is the fact that there is no major pedagogical treatise on the subject of military ethics. As a consequence, the military has had a difficult time stimulating an ethical resurgence from within. In simple terms, the American military has had great difficulty in dealing with its ethical problems because it doesn't know how.

To attempt a treatise on military ethics and to suggest some pedagogical aproaches to its use within the context of the military profession is the primary purpose of this work. This book is a treatise on military ethics which seeks to forge those precepts and obligations that the military professional must accept and observe if he is to be an ethical soldier. It is an attempt to provide a clear philosophical basis for these precepts and obligations. It is not sufficient that a man observe an ethical code; to be truly ethical, to act ethically, and to exercise ethical judgment, he must know *why* certain things are right and wrong, and why he clings to certain precepts. Finally, this treatise offers a code of military ethics and suggests ways of teaching it within the military environment.

Perhaps this book is unique in that no complete treatise on the subject of military ethics has yet been authored by an American. This is probably the case because in a society ruled by self-interest, the fundamental cultural strains of the country—Smith's economics, Social Darwinism, Madisonian politics, and ethical egoism—define such a treatise as redundant or pointless. Certainly, it is historically true that ethics has received little emphasis in any discipline in the United States. Events over the last two decades may have forced a change in these conditions. Perhaps it is time.

In this increasingly complex society, the soldier must have firm ethical moorings. If not, he risks being overwhelmed by the social and organizational forces that restrict his intellect, his will, and his freedom by reducing him to the instrumentality of another's will. If he is unclear in his values, he will seek the ultimate paradox. He will seek to escape from freedom because it is too difficult to deal with it, and in an effort to be free he will most certainly ensure his own enslavement.

Notes

1. For an analysis of these problems as they affected the United States Army during the Vietnam War, see Richard A. Gabriel and Paul L. Savage, *Crisis in Command: Mismanagement in the Army* (New York: Hill & Wang, 1978), Chapter 2.

2. Ibid.

3. Ibid.

4. Douglas Kinnard, *The War Managers* (Hanover, N.H.: University Press of New England, 1977), p. 75.

5. William Hauser, "Armies and Societies: Three Case Studies," *Military Review* 52 (July 1972): 4-5.

6. For an analysis of the AVF and its problems, see Richard A. Gabriel, "About Face on the Draft," *America* (February 9, 1980): 95-97.

7. Ibid.

8. Or so Napoleon thought when he said that "the moral is to the physical as three is to one."

9. A good example of this argument as it applies to the British can be found in Barrie Paskins and Michael Dockrill, *The Ethics of War* (University of Minnesota Press, 1979).

10. These data are based on a selection of four journals for each profession and an actual count of the number of articles dealing with ethics that appeared in each between 1969 through 1979.

11. The Harvard study is detailed in Steven N. Brenner and Earl A. Molander, "Is the Ethics of Business Changing," *Harvard Business Review* (January-February 1977): 57-71.

12. The data cited are drawn from the Harvard study cited earlier and the *Army War College Study of Military Professionalism* (Carlisle Barracks, Penn.: U.S. Army War College, 1971).

13. Francis B. Galligan, *Military Professionalism and Ethics* (Newport, R.I.: Naval War College, 1979), pp. 26-32. It is, of course, an open question as to whether the shift in behavioral standards is a sign of a more permissive or more tolerant society. What seems beyond dispute is that a wide range of socially acceptable behavior would be clearly disruptive of the morale and discipline required of cohesive battle units if allowed to operate within the military.

14. Richard A. Gabriel, "What the Army Learned from Business," *New York Times*, April 15, 1979, p. 52.

15. Kermit D. Johnson, "Ethical Issues of Military Leadership," *Parameters* 4 (1974): 35-39.

16. Ibid.

17. This term is taken from Galligan, *Professionalism and Ethics*, p. 52.

18. Basic reading for this section would include Adam Smith's *The Wealth of Nations*, James Madison's *Federalist* No. 10 of the *Federalist Papers*, and Herbert Spencer's *Social Statics*.

19. Derek C. Bok, "Can Ethics Be Taught," *Change* (October 1976): 30.

20. Edward L. Katzenbach, "The Demotion of Professionalism at the War Colleges," *U.S. Naval Institute Proceedings* 91, No. 3 (March 1965): 34-41.

2 THE NATURE OF MILITARY ETHICS

If military men are going to be held responsible for conducting themselves in an ethical manner, they must gain an understanding of the nature of military ethics as part of the general discipline of ethics. Since most members of the military profession have no training at all in ethical reasoning, they are often unclear about what ethics is and the obligations it entails. Even the requirement that soldiers observe the precepts of a code of ethics will have little meaning if they do not understand why certain precepts must be obeyed and others rejected. While it is often felt that philosophy is the antithesis of pragmatic action, no member of the military profession, especially in the officer corps, can be expected to make pragmatic ethical decisions without an understanding of the philosophical foundations of those decisions. Ethical action in the military profession presupposes and requires that officers and men grasp at least the rudimentary elements of what ethics is all about. The debate over the role of ethics in the military is centered more upon the failure to understand the fundamentals of ethical action than upon real issues. This chapter hopes to provide an understanding of the nature of ethics and some of its more important elements as they apply to the military profession.

An Ethical Perspective

Military ethics forms the core for a profession that is engaged in a very special task sometimes requiring the sacrifice of human life as well as the deliberate taking of human life. Given the role of the soldier, it is clear that some code of values is necessary to give a human and humane dimension to the soldier's awesome tasks and responsibilities. Without ethics, the soldier can easily slip into the moral morass of Eichmannism —the value-free technician who applies his skills in a moral vacuum simply because they are ordered by the state. He can also lose sight of his special obligations and come to regard his personal goals and needs as the sole determinant of right and wrong. In this case he falls into the trap of ethical egoism: he becomes an entrepreneur who uses his position of special trust and confidence to enhance his own career as the highest operant value. Either road leads to ethical ruin, to say nothing of military ineffectiveness. Without a strong ethical compass, the soldier not only can become a destroyer of humanity, but, under the stress of battle, he may also collapse psychologically, for he can lose sight of the reasons why he is practicing his profession.

Clearly, the military must operate in the context of the larger society it serves. However, members of the profession must carve from that society a place of action in which the values of the profession are the predominant moral influence. The ethical precepts of a profession which are expected to guide the behavior of its members are often different from and even, in some instances, antithetical to larger social values. Society is highly role-differentiated. A military officer may at the same time also be a Rotarian, a father, the head of a family, a member of a church, and active in community organizations of various types. He thus acts within the context of several roles in different social settings. It would be incorrect to expect ethical precepts to operate in ignorance of social role-differentiation. This differentiation, especially in modern post-industrial societies which tend to be highly complex, fragmented, and compartmentalized, affects ethics in that standards of ethical behavior apply to different roles and demand different types of action.

This is not to suggest that social roles may ever condone or require unethical behavior. Although members of a given profession may be expected to behave differently in their roles as professionals than they do in their other roles, it is clear that no professional ethic can ever condone evil in itself. No ethical proposition can ever be *malum in se*, that is, evil in itself. Thus, members of the military are expected to observe a particular set of ethical precepts that are most relevant to behavior in that profession. The fact is when ethics are regarded as statements of obligations that one ought to observe in certain circumstances, then the circumstances in which one has to decide what to do and what constitutes ethical behavior have very much to do with what those obligations are in the first place.

It is also clear that different professions require different precepts to guide ethical behavior. A list of precepts delineating what one ought to do as a military professional would differ significantly from a list of precepts of what one ought to do as a member of the medical or legal profession. Now this position implies that what one "ought to do" is strongly conditioned by the profession in which one claims special membership. That is, after all, one of the primary definitions of a profession—that it requires special kinds of behavior not generally shared by the larger society or, at best, shared only tangentially.

Ethics is therefore a social enterprise. Professional ethics is forged in and applies to social settings. It is not the mere invention of the individual, and it exists prior to him in much the same way as a language exists prior to learning it.[1] The individual acquires a sense of what he ought to do, namely, a sense of ethics, when he gains membership and participates in the profession. Furthermore, the ethic of the profession lives on, even if the participant dies. This suggests that the profession itself has an obligation to preserve certain values even if, from time to time, some of its members fail to live up to them. Thus, ethics has social origins and social functions, namely, to guide individual and group behavior. And ethics has social sanctions, the most obvious being the expulsion from the profession of individuals who fail to observe its standards. In a very real sense, ethics constitutes promises made in a social setting to observe certain kinds of behavior in certain kinds of circmstances. None of this implies that ethics is made up of purely relative standards or that ethics is "situational." It means simply that ethics is a human invention and a statement as to what constitutes proper behavior within a profession, and that standards of ethical behavior are rooted essentially in social enterprise and the dynamics of the profession itself. More will be said in this regard when discussing the origin of obligations. Suffice it to say here that different professions require different ethical obligations from their members and that the construction of ethical obligations by a profession represents a carving out of a kind of ethical space from the larger society.

Military ethics consists of a set of propositions specifying the actions of its members as right or wrong. It deals specifically with those values and expected rules of the profession that are appropriate to actions taken within the military environment. Accordingly, when speaking of military ethics one is addressing the problem of living an ethical life within the context of the demands and obligations levied upon the membership of the military profession. To be a good professional is not the equivalent of living an ethical life.

For the soldier military ethics is only one part of his total ethical self. He will also be subject to the ethical claims made upon him by the other roles he occupies. The values and obligations of the doctor are different from those of the soldier, as they are again for the lawyer. Ethics specify obligations relative to the conditions under which they are expected to be observed. Such obligations *may* have some absolute value but this is not necessarily the case although it is clear that an ethical precept can never be intrinsically evil as such. Thus, an ethical obligation need not be absolutely good in all instances, but no ethical precept may be *malum in se*. That which is absolutely wrong in a larger ethical context, such as murdering children or shooting civilians, cannot be made ethically right simply because it occurs within the context of a profession. Those things which men regard as *malum in se* remain such and no amount of professional loyalty will make them right.

The Nature of Ethics

Before turning specifically to military ethics, it is worthwhile to examine the subject of ethics in general because much of what can be said of ethics per se clearly applies to military ethics. And if one is going to examine ethics, the thoughts of Socrates on the subject are most instructive. In the *Crito*, Socrates lays down some basic distinctions which will constitute points of departure for the discussion of ethics throughout this book. These distinctions are basic assumptions about the nature of ethics and clarify some of the difficulties attendant to the debate on military ethics that has been carried on within the profession for the last decade.

In the *Crito* Socrates states that ethical questions and decisions are best approached through the use of reason. Ethics is assumed to apply to men, and that which distinguishes man from other creatures is the possession of an intellect. However one defines intellect, it is clear that men commonly attribute to other men a quality called reason, the ability to explain why men do what they do as well as the capability to make judgments as to whether what they do is good or bad, right or wrong, ethical or unethical. When one is dealing with ethics, one is involved in the process of moral reasoning. Ethics has to do with the power of man to reason, and, as such, ethical questions are best approached through the use of reason and not emotions.

Another Socratic distinction that is crucial to understanding ethics is that one cannot find answers to ethical dilemmas by simple reference to what others think. The ethical man must above all remain the agent of his own fate. He must bring to bear his own reasoning powers, and he must shoulder ethical responsibility for what he chooses to do in given circumstances. The notion that a soldier was "only following orders" or that an individual acted a certain way only because others did is *never* an acceptable mode of ethical reasoning. There must be reasons why one ought to do some things and reasons why one ought not to do others. A man who observes a code without knowing these reasons and without understanding them cannot be said to be acting ethically. He can only be said to be obeying a set of precepts that he does not understand.

A basic ethical principle which runs through virtually all treatments of ethics is that one ought always to do what is morally right and never what is morally wrong. Or, as it is often put, one ought to do what one ought to do. While it appears that these two propositions are two sides of the same coin, they actually are not. Often, it is much easier to know what one ought not to do than to know what one ought to do. The point is that the constant tension between having to do what is right and not having to do what is wrong constitutes a central dynamic of ethical action. It drives home the fact that ethics are by nature prescriptive and proscriptive, that they require some things and prohibit others. It is the task of the human being as a moral agent, as it is the task of the soldier as a member of his profession, through the use of his reason and ethical principles, to decide and to understand why under a given set of circumstances he must do something or not do something. It is in the solution to such problems that ethical action is found.

Socrates identifies something called working ethics, which he defines as a pattern of moral reasoning that helps the individual determine what he ought to do and what he ought not to do in certain circumstances. This moral reasoning occurs with reference to certain specified general ethical precepts. But ethical action necessarily involves the agent in making concrete choices; thus, one "concretizes" general ethical principles to the specific. If necesary, one makes choices about which obligations take precedence over others. It is a fundamental fact of ethical action that any given ethical code may confront the individual in any given set of circumstance with a choice between conflicting obligations. Under these circumstances, ethical choice consists of the ability to choose one obligation over the other in circumstances that prohibit one from doing both. The point that Socrates made two thousand years ago and the point worth emphasizing here is that ethics has to do with the use of reason in *choosing* what one ought to do. This reason is combined with prudential judgment, which helps the individual to determine how his ethical obligations apply in particular circumstances.

Clearly, ethics involves choice and judgment. As regards judgment, men make two kinds of ethically related judgments: those about moral obligations, that is, about what people ought to do or ought not to do; and those about moral values, that is, about what things are seen as good in themselves or evil in themselves. Judgments in an ethical sense apply not only to the individuals who make them but to others as well. Making judgments about how one ought to act is important because in so doing we imply that this ethical judgment ought to be made by other men in the same circumstances. Thus, we not only make judgments about how *we* ought to act but also about how *others* should act. As moral agents, ethics is not only a guide to our actions, but also serves as the basis for judging how other members of our profession should act. From the perspective of ethics, if a soldier would be ethical he must realize that he is not only an actor but also a spectator, adviser, judge, and critic in the ethical drama. A code of ethics in any profession must allow its members to perform all of these roles. Although an ethical judgment is intensely personal in terms of the consequences for the individual making it, once it is made there must be the clear implication that what a man may have done in particular circumstances is justified as appropriate for other men to do in the same circumstances.

A Definition of Ethics

It seems appropriate to offer a definition of ethics that will serve as an analytical schematic for searching further into the nature of military ethics. The object of the search is to point out those kinds of obligations, difficulties, and solutions that generally arise in an understanding of ethics in general and its applications to the military profession.

The Nature of Military Ethics   29

When talking about the meaning of ethics, at least two types of activity come to mind. First, ethics has to do with the way one thinks about moral questions. From this perspective, ethics involves moral reasoning and is a branch of moral philosophy. It represents a mental application of knowledge about questions that arise in the mind of the actor concerning moral dilemmas. More to the point of military ethics is a view of ethics which sees at its core the observance and undertaking of ethical actions. By this, one means living up to the obligations and precepts expressed in a code of ethics and observing the obligation to make choices when ethical obligations conflict. Ethics, then, has to do with the translation into human affairs of ethical precepts requiring or prohibiting certain kinds of action.

In this view, ethics has to do with observing obligations and knowing why it is one observes them. It is worth pointing out again that the mere compliance with specified precepts or obligations is not an ethical act unless one is aware that one is observing an obligation and knows why. At the very heart of the study of ethics are the concepts of obligation, knowledge, moral reasoning, judgment, and, of course, the necessity to choose which obligations take precedence in a given set of circumstances when ethical precepts conflict.

A good working definition of ethics is the art of observing those moral obligations that are appropriate to a person's roles in the social order. Military ethics can be defined as the art of observing those moral obligations and precepts that are appropriate to a person's role within the military profession. Again we confront the proposition that ethical obligations are, if not totally role-specific, certainly most relevant to the conditions in which the individual is likely to be asked to observe them. Membership in a profession is likely to present the individual with specific kinds of circumstantial difficulties in the application of ethical principles that inevitably must be taken into account in the ethical equation. In this view, ethics becomes a form of moral promise-keeping.

Underlying any code of professional ethics is the idea of obligation, especially of moral obligation. Moral obligations acquire a special character insofar as they are recognized to involve grave questions of right and wrong. Thus, an obligation to be prompt, for example, is not regarded as possessing the same moral gravity as ensuring that one's men are not needlessly squandered in combat. Ethics involves keeping not just promises but moral promises, those promises perceived to be of some moment when they are not kept. Not every obligation is a moral one, not every obligation is of equal moment, and the weight of obligations may change relative to the circumstances in which they must be observed.

Obligations

Much of the confusion surrounding military ethics has to do with the failure to understand the nature of obligations and what they imply. Obligation has to do with *actions*, not behavior. By focusing only on behavior, one can easily fall into the behaviorist trap which considers the total of man's ethical action to be evident only in his behavior. From this point of view, one is only what one does. Moreover, a simplistic focus on behavior would make no moral distinction among men who acted out of tradition, habit, coercion, or even blind fear. Observing ethical obligations clearly involves something that goes beyond mere behavior; it goes beyond behavior to action. Action implies the freedom to obey, that is, it implies a certain freedom from overt coercion as well as the more subtle kind that may lurk in habit or tradition. Action also implies the freedom to disobey. Accordingly, when the ethicist talks about obligations, he is ultimately speaking about *willful* acts, and not acts rooted in coercion, habit, or tradition. Observance of an ethical precept out of habit or coercion is not really an act of obligation. Obligation implies the ability not to do what one ought to do.

Along the same lines obligations imply *ability*. A fundamental proposition of any ethical theory and the moral judgments that it makes about men's actions is summarized in the axiom "ought implies can." There can be no basis for judging actions as moral or immoral when the ability to perform, or indeed not to perform, the judged action is absent. Obligations are morally binding only when it is possible to execute them. One could never, for example, have an obligation to jump over the moon because it is not possible to fulfill such an obligation. If one is going to impose a number of ethical obligations on members of a profession, the members must be aware of them and have the clear ability to perform them.

Obligations are also concerned with the activities of *reason*. If an obligation involves a willful action, it is clearly based in the psychological activity men have come to call reason. Accordingly, one must consciously know when one is obliged to obey and when one is fulfilling an obligation. If an individual is not rationally aware that he has an

The Nature of Military Ethics   31

obligation, then no obligation can be said to exist in any moral sense. This does not mean that the community cannot impose sanctions against him, and there are frequent instances when it does so. It only means that, from the perspective of moral judgment, it is difficult to argue that one should be held responsible for not doing what one ought to do when one is unaware of what one ought to have done. This is one of the stronger arguments for the formulation of an ethical code for the military profession. Soldiers cannot truly be held ethically responsible for obligations unless they are aware of them. A formalized code of military ethics is one of the surer ways of informing members of the profession of their ethical obligations as professionals.

The argument that obligation involves reason implies another position: not only must an individual be aware that he is fulfilling an ethical obligation but, equally important, he must also know *why* he is doing it. An individual who complies with an ethical precept without knowing why he must comply is not truly carrying out an obligation; he is merely exhibiting obedience. This raises an interesting point about ethical codes. If it is true that men have to know why they observe obligations, it is equally true that they are not likely to readily develop reasons as to why they ought to comply until they develop and are aware of a code specifying a set of ethical precepts. Merely complying with an ethical code does not constitute ethical action. Rather, the individual must understand that an obligation is present and that he must observe it because there are sound reasons for doing do. Anything short of this obviates ethical judgment and substitutes mere obedience for ethical decision-making.

Ethical action requires *alternatives*. Fulfilling an obligation implies that there are alternative courses of action which the actor has examined and has chosen not to undertake. If there are no alternatives, the individual cannot in any meaningful sense be said to be executing an obligation or behaving ethically. Again, he can only be said to be performing an act of obedience. An obligation is operative only insofar as alternative courses or obligations exist, are recognized to exist, and are not chosen in favor of executing the obligatory precept.

A final element of obligation has to do with choice. The availability of alternatives clearly implies a *choice of actions*. If an obligation is truly an obligation, alternative claims must be recognized and rejected in favor of observing the ethical precept. This necessity for choice not only implies the recognition of alternative claims in terms of other obligations, but also recognizes that the presence of coercive forces may present one with a serious dilemma. Thus, men cannot ethically do some things even if they are threatened with coercion or loss of life. If, for example, one recognizes the principle that executing innocent children is *malum in se*, then the soldier who is ordered to kill children under the penalty of his own death and does so is not acting ethically. If he recognizes that the initial premise not to kill children is what is right in these circumstances, the fact that there is a coercive force operant in the ethical equation merely constitutes an alternative claim. If he is to act ethically, he must reject that alternative claim and refuse to carry out the order to execute children. If he carries out that order, then the fact that there was an alternative claim in the form of a threat to his own life does not remove his ethical responsibility.

The essence of ethical judgment and of observing ethical obligations consists in the recognition that men must often choose among competing obligations. When one obligation is clearly superior to another, the necessity for choice does not present much of a problem, all other things equal. A true ethical dilemma exists when an individual is confronted with two equally compelling ethical obligations under circumstances that do not allow him to carry out both. In this case, he must make an ethical judgment which implies that his action is guided by his reason and an understanding in his own mind as to why he has chosen one course of action over another. To remove the element of choice from ethical decision-making and to attempt to substitute blind obedience is to reduce the obligatory nature of ethics and to transform all ethics into mere questions of who obeys whom, that is to say, into questions of power.

The elements discussed above are essential conditions for an obligation to be present and binding. Grave ethical dilemmas do not in an empirical sense constitute a source of constant concern for most individuals in most professions on a day-to-day basis, although they may arise more frequently in the military. Ethical decisions are not concerned with trivialities and rarely concern the routine operations of our lives where, for the most part, ethical issues are not raised very often. Most men go through most days in most professions most of their lives not having to confront serious ethical challenges because serious conflicts of moral obligation simply do not arise. Moreover, man's tendency to build highly complex, socially differentiated structures also removes the loci of decision-making for many of the serious issues of his life, leaving him to deal with trivialities. Not every disagreement is an ethical issue, and not every unclear course of action involves an ethical choice, nor is every refusal to comply with instructions necessarily based in ethical considerations. It must be remembered that ethical considerations involve moral obligations. Thus, in most instances the problem of solving an ethical dilemma does not arise. In normal day-to-day career activities, most individuals comply with existing standards because they do not appear to them to raise ethical questions and, most often, they do not. There is, of course, the danger that existing norms can institutionalize unethical practices or, at the extreme, degenerate into a "banality of evil."

Observing obligations does not always mean taking action in compliance with the norms of the profession. If one imparts a moral quality to the notion of obligation, the course of action selected must be undertaken because it is seen as morally right. If so, a moral course of action can be either to obey or disobey. One of the elements of moral judgment is the willingness to make decisions among competing moral claims when to observe one is to obviate the other. To raise the question of ethical obligation is to be aware that the sword cuts in both directions. Ethical obligations obligate. At their most basic, they obligate one to make choices among competing obligations. From this perspective, the professional soldier must be aware that he has an obligation to observe certain codes and that at some point he may well have an obligation to violate some precepts of those codes. The necessity for choice is simply inescapable in human affairs.

Sources of Obligations

How do men acquire obligations, and how they come to feel obligated? The answers to these questions are central to the development of any notion of professional ethics since ethical men ought to act essentially in terms of the obligations they acquire in a social environment. As Socrates knew, ethics is closely bound up with the social process. It is very difficult to talk about obligations without inevitably involving the social processes through which they are acquired.

In discussing the sources of obligation, it must be noted that we are dealing with human beings with all their strengths and all their weaknesses. Human beings appear to be largely, if not totally, a product of social environment. Accordingly, the process of acquiring obligations and the way in which they become recognized may be said to be social processes. As a consequence, it is logical to suspect that no one is born with any specific obligations. While it may be argued that men have some obligations to other members of their kind as a result of being men and, therefore, in that sense are born with certain obligations, the point remains that the acquisition of a sense of professional ethics can in no meaningful sense be said to be innate. The reason is that professional ethics seeks to locate and define specific obligations as they apply to the circumstances that are most likely to confront a moral agent as he acts as a member of a profession. Since human beings *qua* human beings are not members of any profession but only become members later on, it is clear that the acquisition of obligations in the sense of recognizing ethical precepts that relate to a profession requires membership in that profession. Hence, the inculcation of professional ethics is closely tied to the social process of group membership.

As a result of social processes, obligation begins with group membership. This is not to suggest that individuals do not feel obligations outside of group membership. Rather, membership in a group seems to make obligations meaningful inasmuch as they require certain kinds of behavior which both the actor and other potential actors who are judging his actions as members of the group recognize as right. Since an ethical judgment implies that one's own actions would be right for other men in similar circumstances, it is clear that the multiplicity of individuals involved in a group provides a key element of ethical action, namely, judging the actions of other members as ethically acceptable or unacceptable. In any case, membership in a group implies certain patterns of behavior or roles which confer expectations on the rest of the group. Simply put, membership in the military profession implies that one will act in a certain way relevant to that profession. This way of acting is likely to be importantly different from the actions of a business corporation, labor union, or even another profession. Membership in a group implies and requires certain ways of acting, as well as a willingness to accept the standards of the group as a yardstick for judging those actions.

Group membership confers certain expectations. Men in the military are expected to act in a different way than lawyers or doctors. Furthermore, all are expected to have different priorities and values relevant to their ethical behavior within their professions for the precise reason that the circumstances under which the soldier is expected to act and the kinds

The Nature of Military Ethics   35

of ethical questions that are likely to arise will be different than for members of other professions. When one joins a group, therefore, one is prepared to accept the obligations of membership in that group.

It may be argued that such expectations are unrealistic and that entering a profession does not confer any obligations since the ultimate obligation for the individual is to pursue what he believes to be right regardless of the precepts of the profession. The difficulty is that membership in any group implies expectations not because they may be ethically right in some cosmic sense, but precisely because without expectations as well as a reasonable chance that these expectations will be carried out, the basis for *any* form of social organization would be immediately destroyed. If the principle that men should do whatever they can to press their own interests were to be universally applied, then lying, cheating, betrayal, and treachery would become the basis for social life. There would then be no basis for social life at all; what would exist would be a war of all against all. The principle that affirms that individuals can lie, cheat, and steal in pursuit of their self-interest cannot be valid insofar as the same individuals who are prepared to condone these activities in their own actions cannot realistically countenance the same behavior for other men with whom they have to deal. To do so is to negate any kind of social intercourse except civil war. Thus, the assumption of certain obligations and the relative assurance that they will be carried out to some degree at least rests at the foundation of any type of social organization. They most certainly rest at the very roots of a profession. It is these expectations that come to constitute obligations. When these expectations are joined with the other characteristics of ethical action, namely, reason, alternatives, judgments and choice, the observance of obligations in an ethical sense commences.

The social proclivity for observing obligations appears, therefore, to result from values and expectations based in group membership. There is a tendency to elevate these expectations into statements of principle or precepts as to what is "right" for members of the profession or group. However, when viewed with a jaundiced eye, commitments to professional precepts in the empirical world are only commitments to other men in groups that affirm certain values and require certain actions and thus levy specific obligations. Apart from groups, it is very difficult to conceive of obligations in anything but an abstract sense. It may be simply mindstuff to expect the monk to discharge the obligations of a monk when there is no group of monks to expect and judge his action. Of course, one can always claim membership in a higher group—say the eternal brotherhood of saved souls—to justify one's actions. In the empirical world, however, it might be suggested that obligations that rest only with one's self and are not rooted in group membership or judgment are not really obligations as much as they are statements of personal preference. Obligations imply that one is fulfilling prior expectations in a manner that a social group in which one claims special membership would approve. It does not imply that one is merely "doing one's own thing." Thus, obligations require the existence of a social group to have ethical meaning.

If obligations are strongly rooted in the social process, a professional soldier may acquire several distinct sets of obligations derived from his various sets of social interactions. Therefore, in some situations one's obligations to one group might well conflict with one's obligations derived from membership in another group. When this happens, the moral dilemma surfaces and the individual must choose which obligation will take precedence and to know why it does so. This is the crux of ethical judgment, and there is no certain way in which one moral obligation can be assessed *a priori* to be greater than another apart from the circumstances in which they must be observed.

Precepts

As observed earlier, ethics has to do with precepts, that is, ethics makes statements as to what men ought to do and ought not to do. All ethics is prescriptive and/or proscriptive. This idea, simple as it is, does not appear to be widely accepted in military schools, at least as is evidenced by their ethics curricula. The current view in these schools is that teaching and inculcating ethics does not involve the specification of precepts that are regarded as obligations. Rather, it is assumed that the fundamental problem in teaching ethics is to offer a range of ethical perspectives to the individual and have him select whatever set he is most comfortable with. This view, of course, suggests that to describe ethics is to learn ethics, and it confuses the teaching of ethics with ethics per se. In teaching ethics one can present a wide range of ethical perspectives as long as the purpose is only intellectual exercise and not the inculcation of ethics. But if one wishes to inculcate a sense of ethics in an individual, one must specify what set of ethical precepts he prefers to others and be able to provide reasons.

The idea that teaching and inculcating ethics involves merely the presentation of a number of ethical perspectives from which the individual may extract those with which he feels most comfortable is called descriptive ethics. Descriptive ethics is by definition a contradiction in terms. All ethics must either proscribe that the individual not do some things or prescribe that he do others. It is this binding quality, this sense of the imperative, that separates ethical obligations from other kinds of rules. The idea that an individual may pick and choose among a series of ethical perspectives simply because he is comfortable with them is a normative paradox. It is especially so in the military profession which with its stated concepts of obligations and responsibilities is set apart from the larger society. If all perspectives of right and wrong are equally acceptable, then there is no basis at all for any ethical judgments. What distinguishes ethics from mere dissertations on the subject, or what distinguishes ethics from teaching ethics, is precisely this prescriptive and proscriptive quality which imposes a number of serious obligations on the individual. As Alvin Toffler has suggested, "If there is no vital center to man that is dynamic and unique that acts in terms of higher standards then all expressions are equally valid."[2] No one who seriously considers the nature of ethics can agree that all expressions of or justifications for human action are equally valid or equally right or equally wrong. To suggest that individuals may validly select any precepts to guide their behavior is to deny the obligatory nature of ethics. At the same time, it is to deny that the center of ethical action is making choices between obligations. In terms of descriptive ethics, the "expressions" of the saint and the assassin are equally valid.

The idea that one can inculcate ethics in the soldier by exposing him to a range of ethical perspectives and allow him to select those he wishes and reject others implies at the very least that ethical obligations need not be specified by the profession, that is, that an ethical code is not required. But to suggest that an ethical code is not required is to negate the very nature of any profession which, by definition, must posit as its *raison d'etre* a set of precepts, values, obligations, and responsibilities which separate it from the larger society and which apply directly to the members of the profession. Moreover, the idea of descriptive ethics implies that ethics cannot (or should not) be taught within an organizational setting, leaving the choice of what obligations are binding totally with the individual. Both of these assumptions run throughout much of what is written about ethics today, and both are fundamentally wrong.

What is encountered here is a curious application of the idea of free enterprise to the discipline of ethics. To deny the need for an ethical code for a profession is to assume that a collectivity of individuals, officers, and soldiers, whose ethical precepts are fundamentally personal ones, will spontaneously generate a sense of ethics that is applicable and acceptable to the professional community as a whole. But precisely the opposite is true. It is possible to have a community ethic which can be used to integrate and socialize individuals to that ethic as the price of passage and membership in the ethical community or profession. It is unrealistic, however, to attempt to produce an ethical community from a random collection of ethical perspectives. Thus, one of the basic failures of thinking about ethics as they apply to the military profession is the tendency to confuse descriptions of ethics with ethics per se.

Descriptive ethics implies that one cannot teach ethics or enforce ethics in an organizational setting precisely because ethics is a highly personal concern. In point of fact, the ethics of a profession is not a highly personal concern at all. The ethics of a profession is a community concern. It is membership in the profession that lends a particular set of obligations their binding quality. Individuals who cannot accept the ethics of the profession must, therefore, remove themselves from it. It is a contradiction to suggest that the military profession will have no code of ethics and still remain a profession or, equally so, that it has a code of ethics from which members may freely select those precepts with which they agree while rejecting others. Both of these conditions negate the very essence of a profession and, indeed, the very essence of ethics.

Art

Ethics is an aspect of human activity which involves judgment as opposed to the scientific application of rules. For this reason, judgment is at the very center of any ethical code. As noted earlier, to carry out the letter of the obligations specified in a code without knowing why does not constitute ethical action, only obedience. The same is true when individuals refuse to make choices among conflicting obligations and merely follow orders that they do not understand. The idea that ethical action involves art as opposed to science reflects the essential difference between the bureaucrat and the soldier.

The question of bureaucratic ethics may constitute a contradiction in terms. The purpose of a bureaucracy is to routinize decision-making and to eliminate judgment. The bureaucrat first and foremost carries out rules formulated by others in which he usually has no part in framing. Moreover, he is not asked to agree or disagree with these rules, only to execute them. Thus, the bureaucrat in the execution of routinized rules is not really engaged in action, the willful carrying out of obligations, as much as he is engaged in behavior, merely executing rules regardless of why he understands they are there. The fundamental task of the bureaucrat in practice is not to observe obligations but to obey existing rules. The purpose of the bureaucrat in following orders is to narrow his accountability so that he can escape responsibility if anything should go amiss. The logical and, perhaps, even legal defense of the bureaucrat when things go wrong is to affirm that he executed the rules that were required by existing regulations. Because he followed orders laid down by his superiors, he would argue that they, not he, must bear responsibility.

Actions taken by bureaucrats are not at all what we generally mean by ethics. Ethics involves the art of judgment. Instead of the bureaucrat's rulebook, the soldier must have a code of ethics. The code itself is general, and the way it will apply in any given set of circumstances is often unclear and depends heavily upon the soldier's judgment. However, while the bureaucrat is engaged in the behavior of merely carrying out rules, the soldier can never be allowed to engage only in the behavior of carrying out the code. The level of generality of the ethical code requires that the soldier be aware of the basis of the code's precepts. It is not enough to execute an obligation. The soldier must engage in ethical action, that is, he must willfully carry out the obligation, and he must know why it binds. The soldier must exercise ethical judgment. He is engaged in the rational action of discerning why one obligation binds more than another in a given set of circumstances and why he must choose one obligation over another. Accordingly, the last refuge of the bureaucrat is to execute rules as a means of escaping responsibility. *The member of the military profession must exercise judgment and can never escape responsiblity for his judgments.* Indeed, one of the major differences between the bureaucrat and the military professions is that the one seeks to escape responsibility whereas the other accepts it. One of the curses of the modern military is the tendency of many to confuse membership in the profession with membership in a bureaucracy, and then to complain that the military seems to have lost its ethical compass. Being a member of a profession is categorically different from being a part of a bureaucratic apparatus. The rules are different, the ethical requirements are different, but most of all the degree to which judgment is required and the extent to which responsibility must be directly assumed are categorically different. The very idea of an "armed bureaucrat" is itself a contradiction.

Responsibility

Members of a profession must exercise judgment, which means they must assume responsibility for choosing some obligations over others. Thus, another important aspect of ethics is the proposition that individuals can never escape responsibility for their ethical choices. One characteristic which distinguishes human beings from other species is rationality. Another is their capacity to make conscious, rational explanatory judgments as to why things are right or wrong. In this sense, all men are ethical agents who are totally responsible for their acts and their consequences. To abandon ethical judgment to another, for one individual to subordinate his ethical judgment to another, is to cease to be human. Certainly, such a person ceases to be an autonomous ethical agent and becomes a tool of another's will. As a defense of unethical action, the idea of absolute obedience to one's superiors can never be applicable to the relationship of one human being to another.

As ethical agents, all men are equal: they can in principle be held equally responsible for observing their ethical obligations. Trusting another's ethical judgment implies an inequality of power and perhaps of knowledge. One can certainly learn from others or trust them in certain circumstances. But an inequality of power or even knowledge cannot replace the responsibility that goes with ethical agency. As a human being, man remains always an ethical agent, and there is no escaping the awesome responsibility to act ethically.

Another reason why men cannot escape ethical responsibility is determined by Western tradition of custom and law which affirms that one must bear responsibility for one's acts. Whether one examines the fifteenth-century doctrine of *respondiat superior* ("let the superior be responsible") or the notion of just war, or even the more recent cases of the My

Lai massacre or the execution of General Yamashita, or, finally, the Nuremberg trials, it is clear that Western society has long held that men cannot escape ethical responsibility for their acts by transferring that responsibility to others. The doctrine of accepting ethical responsibility was formally enshrined in the United States military profession as early as 1863 in General Order number 100 of the United States Army Field Manual: "Men who take up arms against another in public war do not cease on this account to be moral beings responsible to one another." Individuals always remain ethically responsible for their actions, for the choices they make among conflicting moral obligations, as well as for the consequences which result from them. To deny that a soldier has ethical responsibility is to negate the very nature of ethics as ethics applies to the military profession.

Members of the military profession acquire and come to feel obligations to the profession and to its values. These obligations involve precepts, that is, specific requirements which individuals must follow if they are to be regarded as ethical. At the same time, an ethical code requires that soldiers make judgments since ethical precepts may sometimes conflict and thus force soldiers to choose among them. Hence, no soldier can escape responsibility for his actions. Ethics, therefore, implies ethical actions in which individuals are held to be ethically responsible for what they do precisely in terms of what they promise to do and not to do. Specifically, soldiers are ethically responsible for observing the code of ethics they agreed to abide by when they acquired special membership in the profession of arms.

For the military, then, ethical action requires the specification of those precepts that are most likely to apply within the military environment. An action is said to be ethical when the consequences of the agent's choice are such that the action attains or approximates the terms of the obligations stated in the code of ethics. The individual is said to have undertaken an ethical act when his choice tends to attain or move toward the attainment of the values expressed in the code. The act is regarded as unethical if that choice tends away from attaining the obligation or even destroys some precepts of the code of ethics at the center of the profession. Members of the profession must set standards of proper military behavior and must observe the standards, being consciously aware of why the obligations bind as they do. When they merely execute the precepts of the code without knowing why, they are engaged only in acts of obedience. Ethical action involving as it does judgment, choice, and responsibility is the antithesis of obedience. Members of the profession of arms must understand that sterile loyalty to a stated code is meaningless unless the precepts are understood and its obligations undertaken willingly. In short, one cannot avoid ethical responsibility by blindly observing a code.

Objections

There are some objections to the development and use of a code of ethical precepts for the military profession. Some of these objections appear to be based on a misunderstanding of the nature of ethical obligations. Nonetheless, since they are raised from time to time, they should at least be identified here.

One objection concerns the concept of ethics as precepts. Such precepts are said to be fundamentally unworkable because no ethical code could ever specify in advance all the circumstances in which a precept would apply. It is true that no ethical code can specify every situation to be covered by specific precepts and that it is impossible to develop a code in which the precepts of the code cannot be made to conflict in some situations. It is very nature of life to be uncertain. However, these difficulties are not substantial enough to support the proposition that one cannot develop a working code of ethics, nor do they negate the value of an ethical code for a profession.

To argue that an ethical code is useless because it cannot specify in advance what ethical action ought to be taken in every possible instance is to confuse ethical codes with a body of laws. This is, as Malham Wakin has pointed out, a distortion of their function. "The immature or unsophisticated frequently narrow their ethical sights to the behavior specifically delineated in a code so that what may have originally been intended as a minimum listing becomes treated as an exhaustive guide for ethical action."[3] It is the nature of laws to address specific instances of behavior and to detail how one must behave in these circumstances in order not to be guilty of illegal behavior. Ethical codes, on the other hand, set forth general principles and statements as to what one ought to do. The interesting distinction between laws and ethical codes is that laws require obedience without any understanding as to why the law necessarily is as it is. Ethics, on the other hand, sets forth general statements about what one ought to do and requires that an individual know why the precepts constitute obligations. Moreover, codes of ethics require the application of judgment in order to decide how a precept applies in given circumstances; laws require no such judgment. Laws are only the dictates of the state; they may or may not have ethical content. Ethics by its very nature addresses the moral content of human action. It is the very nature of ethics to set forth principles of action rather than to specify in advance how a precept will apply in a specific set of circumstances. Only ethical judgment with its attendant necessity for choice can determine the application of ethical precepts.

The argument that ethical precepts are useless because they do not spell out the specifics of ethical conduct in all circumstances misses the point. The crux of ethical action is to choose one obligation over another when one cannot observe both. To suggest that the precepts of a code never conflict would eliminate the conflicts between obligations that require ethical judgment. If it were possible to specify in advance all applications of an ethical precept, the need for ethical judgment would be eliminated altogether, as would any uncertainty in human affairs. Ethical judgment would be replaced by a handbook of rules. One would again confuse obligation with obedience and action with behavior. Without uncertainty in human actions, ethics becomes meaningless.

Another argument suggests that the code becomes useless because it does not provide guidelines as to what precept should be observed. To suggest that ethical codes are useless because some of their precepts may conflict in some circumstances is to misunderstand the nature of both obligation and ethical choice. With regard to conflicting precepts, there is a distinction that may be drawn between prima facie duties and actual duties. An actual duty is what an individual ought to do in a particular situation, while a prima facie duty is what the individual ought to do if no other considerations interfered.[4] It is what normally would be an actual duty if no other moral considerations intervened. As to the precepts of a code, they can be seen to be statements of prima facie duties—those things that ought to be done if no other circumstances or moral considerations were involved. A code cannot contain within it a contradiction of its prima facie duties, for such duties constitute ethical precepts that one ought always to try to observe. But given the conditions under which human beings must ethically act, one ethical precept may be outweighed by other obligations, or indeed by other prima facie duties. It must be remembered that the ethical code for a profession is never the sum total of the individual's ethical obligations, any more than his life within the profession is the total of his ethical life. Accordingly, an ethical code for a profession pertains only to the activities normally encountered in circumstances relevant to that profession. Individuals may perceive contradictions between the requirements of their profession and the obligations they have acquired through membership in other social groups such as the church or family. In these instances, one is left with the very human problem of choice. And one must choose.

The argument that the precepts of a code will sometimes conflict and, therefore, are useless in deciding what ethical action ought to be is not valid, for ethical precepts within an ethical code do not conflict. An ethical code cannot require two obligations that are contradictory. What confuses the individual as to which obligation is more compelling are the circumstances in which precepts may have to be applied. The tension in the mind of the individual rather than the precepts of the code creates contradictions. In any case, it is the very essence of ethical action to make choices. One can only ethically fail to observe an obligation when it is in conflict with another judged to be more compelling and when both obligations cannot be observed at the same time:

> The ground rules for violating a moral commitment are subject to the same considerations as are relevant to other moral contexts. We are justified in violating one of our moral obligations just when that obligation is in conflict with another, higher obligation and the circumstances are such that we cannot fulfill both.[5]

While circumstances may require individuals to choose among precepts of the code, this does not mean that the precepts of the code are not valid obligations. Moreover, because circumstances present the agent with a conflict, the original precepts of the code cannot be said to have no worth or not to be instructive as to how one ought to act. A conflict between ethical precepts of the same code arises out of circumstances and only demonstrates what we have said from the very beginning: that ethical agents must make choices and that choice and judgment are at the center of ethical action.

Conflict can be regarded as a necessary expression of ethical action if the application of ethical judgment is to be possible. To argue that conflicting precepts negate the value of an ethical code is to misunderstand the nature of ethical judgment. Ethical judgment involves uncertainty; it requires the application of an ethical precept in circumstances that compel choices. It is nonsense to try to establish a code of ethics which breeds out the need for ethical judgment. If such a code could be developed, it would not be an ethical code at all as much as it would be a handbook of bureaucratic rules. It would not require human beings as we know them, men with free will capable of choice, but mere obedient automatons to carry it out.

As long as human beings remain what they are, conflicts between obligations will arise. This does not mean that one ought not to clearly delineate what one's obligations are in a profession or in one's life. Nor does it mean that all obligations can be observed equally well under all circumstances. It most certainly does not imply that obligations are meaningless because they are difficult to observe or that ethical precepts are not worth stating because they create difficulties of choice in given circumstances. Simply put, it is difficult to see how anyone can act ethically without first being aware of a set of ethical precepts. Without a code of ethical precepts, it is unrealistic to expect individuals to act ethically. The fact that the precepts of a code may be made to conflict by circumstances says less about the value of ethical precepts than it does about the nature of ethical action and the conditions under which it must take place.

One issue which students of military ethics often raise is whether the stated ethical precepts of the military profession would constitute absolute or relative values. If they are absolutes, so the argument runs, they are precepts which individuals must apply the same way in all circumstances at all points in history. Since few, if any, such precepts can be readily identified, the argument suggests that if military ethics involves absolutes they become impossible as guidelines for human action in an empirical world. On the other hand, some suggest that military ethics consists essentially of statements of only relative obligations. It is argued that the precepts of military ethics are not absolute statements of right and wrong, and are changed and focused essentially by circumstance. To some who see ethics in this light, ethical precepts become meaningless as guides to ethical action since one can never know which precept will apply or how it will apply. From the perspective of absolute ethics it is argued, therefore, that military ethics is meaningless because it is impossible to observe, while from the view of relative ethics it is argued that it is equally meaningless because it does not provide solid guides to its application in varying circumstances.

The argument between relative and absolute ethics demonstrates a basic confusion about the nature of ethics. In the first instance, whether or not ethical precepts of the military profession are to be regarded as absolutes or relatives, at least one point should be clear: ethical precepts have meaning only within the circumstances in which they must be observed. Thus, the argument for relative ethics is really not a valid position to begin with. To suggest that ethics somehow becomes less valid or is of less worth because it applies to empirical circumstances is to miss the point of what ethics is all about. Ethical precepts may have an intrinsic value in that they can never be statements that are *malum in se*. They also have value in a second sense, in that they work to actualize the values codified in the precepts themselves precisely in terms of the circumstances under which a human being will realistically have to pursue them. This is why the ethics of the military profession is different from the ethics of the legal profession. Thus, if one is going to talk about the tension between absolutes and relatives as they apply to the precepts of military ethics, it should be understood from the beginning that the mere fact that ethical precepts must be applied in empirical circumstances and that these circumstances have a bearing upon how these precepts will apply does not change the obligatory character of ethical precepts. Nor does it negate their value.

The debate between absolute and relative ethics is often joined by pointing up the tension between what has been called an *ethics of absolute ends* and an *ethics of responsibility*.[6] As regards an ethics of absolute ends, the central proposition affirms that nothing is ethically valid except adherence to a set of absolute values. Obligations to absolute values logically permit no modification by empirical circumstances. One is required to observe certain ethical precepts regardless of the circumstances in which they occur. What is most important from an ethical perspective is the ethical attitude of the agent; indeed, this is the only concern. If the individual's *intentions* are good, then the consequences of his actions are, in a strict sense, not relevant to the ethical quality of the act. This perspective, termed *Gesinnungsethik*, is not acceptable as the basis for an ethic of the military profession. It is unacceptable because, while it is necessary to examine the motives of an individual in assessing the ethical quality of any act, those motives cannot be the sole factor in judging whether or not an act is ethical. In some instances, even individuals pursuing the noblest of goals can precipitate disastrous consequences. Ethical action cannot consist solely in doing something without regard for its consequences. Indeed, it is even a doctrine of ethical irresponsibility. If one accepts the dictum of *homo mensura*, that man is the measure of all things, that what we do to our fellow men is important in assessing the ethical consequences and character of our actions, then it is clear in a very common sense way that consequences count. Men make judgments about the ethical quality of an act in terms of what the individual tried to do and how valuable that may be. But they reserve a more important place for assessing the consequences of an individual's acts. The idea that a series of absolute values can and ought to be pursued in an empirical vacuum makes no sense, for men can only act in empirical circumstances. Moreover, the implication that in the pursuit of ethical ends one can freely generate frightening consequences is tantamount to the abandonment of ethical responsibility. One cannot affirm that a code of military ethics resides totally in an ethics of absolute ends without risking an ethical and empirical disaster.

In contrast to the ethics of absolute ends is the ethics which is essentially rooted in the responsibility of the actor himself—the ethics of responsibility or *Verantwortungsethik*.[7] The ethics of responsibility presumes that the individual does not adhere to an absolute set of values or, indeed, even attempt to pursue good intentions. Rather, the individual tries to act in such a way as to effect the most humane consequences of his actions. The standard of ethical judgment is the consequence resulting from one's actions. While bad consequences can make a well-intentioned action unethical or immoral, good intentions do not in themselves constitute an ethical act. For an act to be ethical, the individual must realize that he is observing an obligation, and he must have knowledge as to why the obligation is worth observing. Thus, from the perspective of an ethics of responsibility, the focus on the consequences of the act is insufficient as a standard of ethical action. It represents a confusion of human action with behavior. Assessments of ethical actions must include an examination not only of the consequences of action but also of the actor's intentions. Neither by itself is sufficient.

Neither the ethics of absolute ends nor the ethics of responsibility is sufficient to support a code of military ethics. The pursuit of absolute ideals as a kind of self-centered ethical task is rejected as potentially devastating for a profession that deals directly in the lives and deaths of large numbers of human beings. At the same time, concern for the consequences of one's actions cannot be seen as the sole determinant of ethical action. Rather, the individual must have an awareness of his obligations and their value as such. Thus, the question of absolute versus relative ethics comes down to the following proposition: for a code of ethics to be valid, it must stipulate precepts which members of a profession hold to be good in themselves as prima facie duties; that is, they have some intrinsic merit that transcends bad application. At the same time, such precepts cannot be isolated from the empirical world in which the members of the profession must act. The precepts of an ethical code must, therefore, also "work" in those circumstances that a member of the profession is most likely to encounter. Ethics that "work" imply that, if an individual acts in terms of the precepts relative to given circumstances, the ideas, values, and goals expressed in the code will tend to be achieved, or at least approximated in the empirical world. For example, if the code requires that an individual do X, then to do X must be regarded as good and it is also possible to do X given the circumstances under which an individual is expected to do it. If the individual does indeed do X in those circumstances, the consequences of that action will be also judged as good, all other things equal. One can never separate the intentions of the individual and the values he seeks to obtain from the consequences that result from his attempt to obtain them.

It has been argued that the precepts of a code of military ethics are in a strict sense neither absolutes nor relatives. What, then, is the nature of these precepts? The answer is that they are philosophical and ethical imperatives. If one applies the principle of universality, then the tendency to see ethical precepts as imperatives of action becomes clearer. The principle of universality holds that the fact that one makes a moral judgment in a particular situation in pursuit of particular ethical goals implicitly commits one to make the same judgment in any similar situation. More importantly, it means that if one judges one's own action to be ethical in a certain set of circumstances, one is committed to render a similar judgment upon the acts of others who act the same way in similar circumstances.[8] Thus, one tends to universalize lessons and judgments into rules of action and codes of ethical precepts.

The idea of an ethical imperative can be pressed even further, as Kant did when he developed the concept of the categorical imperative. Kant suggested that a categorical imperative was such that an individual was to "act only on that maxim which you can at the same time will to be a universal law."[9] The concept of an ethical imperative is a valid and useful one. The nub of the concept is that, while certain codes of ethics and ethical judgments are made relative to circumstances, they can be universalized into propositions and precepts that serve as standards of judgment for other human actions that occur in similar circumstances. Thus, an ethical imperative is what men would choose if they saw clearly, thought rationally, and acted disinterestedly and benevolently.

There are, then, few truly absolute obligations—those that apply in the same way to all human beings at all times regardless of circumstances. When we talk about ethical imperatives, we mean ethical precepts that hold for all human beings who must act in the same circumstances. Absolute precepts are seen to hold regardless of circumstances, while ethical imperatives depend upon the circumstances in which they must apply.

Ethical precepts are, therefore, instructions as to how one ought to act under certain circumstances. There is also a tendency to universalize them. If one judges an act to be ethically good, one is implying that one can universalize that precept so that other soldiers who act the same way in the same circumstances would be entitled to the same positive judgment. One could validly raise a precept to the level of a maxim or law in the Kantian sense and develop a series of "universal" imperatives. All the concept requires is that one be prepared to explain and justify one's actions relative to extant circumstances, not only for oneself but also for other individuals confronted with the same choices under the same circumstances. The universalization of an ethical precept is not just a lexical one. One does not codify simply to codify. Rather, one "elevates" a code of precepts requiring certain actions in order to extend one's sense of ethics to other men. In terms of a profession, the sense of ethics of the profession is extended to its membership through an ethical code.

While the application of ethical precepts depends upon circumstances, the precepts themselves are imperatives in their own right: they are judged as good if all individuals were to observe them. They receive additional worth in human affairs by their applications and the judgments that result from them. Yet, the precepts should have sufficient value in themselves, so that the consequences of their application can be judged in terms of how their application approximates their realization. One of the criteria of judging ethical actions is whether or not the consequences of one's actions tend in fact to achieve the values specified in the precepts. If they do, one might then surmise that the act was an ethical one, all other things equal. The point is that the assessment of consequences is tied to the internal value of the principle as specified.

From the perspective offered here, it can be said that military ethics consists of a set of precepts that have been raised to the level of ethical imperatives in that they affirm that all human beings who find themselves in the same circumstances as soldiers within the profession of arms are justified in acting the same way. The tension between absolute ethics and relative ethics is, in the proper context, not so much a tension as a confusion in conceptualization. Absolute values can never be pursued without regard for consequences, nor can human actions be judged in terms of consequences without regard for the intentions of the actor or the values contained within the ethical precepts that he has tried to attain. An assessment of both values and consequences must be drawn together into a set of ethical imperatives.

Situational Ethics

The confusion surrounding the issue of absolute and relative ethics often manifests itself in the problem of situational ethics. Indeed, much of the debate about whether or not military ethics should be situational ethics is irrelevant because it is confused as to its very terms. Situational ethics is not the simplistic notion that any decision that takes into account the circumstances or situation in which the soldier must act negates the value of ethical precepts he should observe. Merely considering the circumstances in which an ethical precept must apply does not constitute situational ethics. As noted earlier, all ethics must consider the circumstances in which the individual is required to act, just as it must consider the conflict of obligations and the consequences that follow from the choice of one over another. To argue that the circumstances of one's actions will to some degree condition the manner in which an ethical precept applies does not constitute a case for situational ethics, any more than it constitutes a case for the dilution of the value of the original ethical precepts.

It has been observed in this discussion that the pursuit of absolute values without regard for their empirical consequences is not a valid ethical position. The notion of *justitia fiat pereat mundus* (let justice be done, even if the world would perish) is not only a very difficult one to observe, but it is also likely to lead to great moral problems in most of its applications. More than that, the notion that the pursuit of absolute values can validly ignore empirical circumstances and consequences misses the point of the debate between situational and absolute ethics. The very nature of ethics involves discovering what one ought to do under particular circumstances. It involves deciding what precepts apply in those circumstances and using judgment in their application. For an individual to consider the circumstances in which an ethical precept applies neither makes the case for situational ethics nor diminishes the value of the precepts that the individual must observe. Ethics and the circumstances of their application are inevitably bound up in ethical action.

What, then, are situational ethics, and why isn't it possible for a code of military ethics to be situational in character? Situational ethics is the ethical theory which affirms that basic judgments about what obligations ought to be observed are always purely particular ones. Ethical obligations are always relative only to the *particular* set of empirical circumstances in which one finds onself *at the moment*. Accordingly, each situation is unique, and the individual must decide what to do precisely on the basis of the information that he has available to him *at the moment*. There is no attempt to make the circumstances relevant to any general ethical precept, or, more importantly, no general ethical precept is relevant to any given set of circumstances. The position of situational ethics is that the individual determines what is ethical exclusively in terms of information and knowledge available to him in the set of circumstances in which he finds himself at any moment. It is, of course, impossible to universalize any principles of ethical behavior as imperative precepts simply because each situation is affirmed to be unique. In a sense, the individual approaches each set of circumstances in which he must act as an ethical *tabula rasa*, lacking any guiding ethical precepts.

The reasons why situational ethics cannot form the basis of a code of military ethics are clear enough. Situational ethics offers no standards whatever about what is ethically acceptable or unacceptable. Apart from a set of particular empirical circumstances, no experience can be generalized into precepts of ethical action since circumstances are affirmed to be unique. In such a situation, one can only act "rightfully" on the basis of information gathered while in the circumstances through intuition, that is, one simply "knows" what is right and wrong without any moral instruction or, as in the case of existentialism, one "decides" what to do and in so doing the individual "becomes what he has done." The problem remains that it is impossible from the perspective of situational ethics to develop any precepts of ethical action that would have any meaning at all outside the totally unique circumstances in which they occur. And since each set of circumstances is unique, the fact that one acted one way in some circumstances does not imply that one ought to act the same way in similar circumstances. In brief, strictly speaking situational ethics is not ethics at all.

Situational ethics offers no way of deciding what one ought to do or what one ought not to do. It is incapable of generating any ethical precepts. Moreover, there is no basis for deciding what is right or wrong since what is judged to be right or wrong is totally and exclusively tied to the circumstances of the moment. Thus, one cannot use one's past experiences to help decide what one ought to do in the future. In addition, situational ethics offers no basis for ethical guidance since in the absence of precepts there is no basis for developing ethical guidance. Finally, situational ethics offers no basis for ethical education since all ethical decisions are totally tied to unique circumstances. There is, then, nothing to teach; education that deals only with singularities is not education at all. If the essence of ethical education involves specifying what ethical precepts one ought to observe while teaching a sense of ethical reasoning that helps one apply precepts in varying situations, then situational ethics cannot offer any basis for moral education since all circumstances are unique.

A final confusion surrounding treatments of military ethics is the argument that one cannot specify the central values of military ethics because to do so is a wasted exercise. The argument holds that all ethical codes are puerile because they can be misapplied by evil men, and the existence of a code of ethics does not guarantee compliance. Of course, the mere promulgation of a code of ethics will not guarantee compliance in all instances, although some empirical studies do indeed show that the promulgation of a code does increase the degree to which ethical behavior is observed within a population.[10] Yet, the fact that individuals fail to observe a code does not negate the value of the code's precepts. Codes state what men ought to do. In a real sense one cannot realistically expect men to observe ethical obligations unless they know what those obligations are. Accordingly, the very first step in trying to inculcate ethical action is to promulgate a clear set of ethical precepts.

In order to halt the violation of ethical precepts, it is important to formulate a standard that tells what is being violated. Unethical behavior can never be corrected without some standard against which to measure behavior as ethical or unethical. The inculcation of ethics as a way of correcting unethical practices requires at least a set of precepts which define the standard of ethical action. The existence of a good code of ethics says nothing about bad practice, and clearly the fact that a code may be violated does not diminish the value of the code as such.

With regard to violations of an ethical code, officers and soldiers must realize that carrying out ethical obligations often involves trying to reconcile conflicting obligations and judging which ones are to be observed and which ones are to be overridden in given circumstances. The fact that one chooses to override an obligation in favor of another does not make the overridden obligation any less of an obligation. To conclude that obligations are valueless when they are not observed and to jettison them because they are less than clear in all circumstances is to abandon *any* sense of ethical commitment whatsoever. In the absence of some code of ethical precepts for the military profession, what guides would there be for ethical behavior for members of the profession? The answer is simply that there would be no guides; thus, ethical codes are required to set standards of professional ethical behavior. That such standards may be violated from time to time says nothing about the value of the code itself. The fact that some individuals violate a code of military ethics is irrelevant to the need and value of the code for the profession.

Conclusions

This chapter seeks to introduce the soldier to some basic ethical concepts and to acquaint him with the difficulties associated with the development of ethical precepts for the military profession. It is hoped that a grasp of fundamental definitions and basic distinctions in the area of ethics and ethical action will help dispel some of the confusion attendant to understanding ethics among military men.

If members of the military are to develop ethical standards, any code of military ethics must achieve three objectives. First, it must make soldiers more capable of recognizing ethical dilemmas that may be involved in their decisions. During the Vietnam War, few soldiers saw the difficulties of that conflict in terms of ethical issues. In peacetime, too, the soldier encounters issues that involve ethical questions, but he may not recognize them as such.

Second, military professionals must be taught to reason carefully about ethical questions. If a soldier does not understand why precepts are binding, it is unlikely that he will ever know how to apply them in changing circumstances. Thus, the mere existence of an ethical code and obedience to it are insufficient to guarantee ethical action. The military professional must become adept at moral reasoning, he must know why he acts, and he must know why he chooses one course of action over another.

The third objective follows from the first two: recognition of ethical dilemmas and the development of moral reasoning will help the soldier clarify his own values. In the end, all men must be responsible to their own consciences for what they believe to be right and wrong. There will always be instances in which their obligations as men *qua* men will conflict with the obligations acquired as members of the military profession. When conflicts become evident, grave choices must be made. The choice might even entail leaving the profession itself. Choices between one's role in the military and other roles can be resolved only when the soldier can clarify his own values. Thus, once again it should be noted that membership in a profession and observing its code do not constitute the sum total of one's ethical being. Nonetheless, establishing certain standards of ethical action within the profession can serve as a stimulus to clarifying one's ethical standards as they apply to other aspects of one's life.

For some, the need for an ethical code for the armed services is not readily apparent. One argument against a code is that the values gained through experience in the larger society to which the military belongs can be transferred directly to military life. Once acquired, these precepts serve very well as ethical standards in the military profession. Others argue that no profession can develop ethical standards which do not, in some instances, impose ethical requirements that are different from those of the larger social order. We next turn to a response to these objections.

Notes

1. William K. Frankena, *Ethics* (Englewood Cliffs, N.J.: Prentice-Hall, 1973), p. 6.

The Nature of Military Ethics 55

2. Alvin Toffler, *Future Shock* (New York: Bantam Books, 1977), p. 17.
3. Malham M. Wakin, "The Ethics of Leadership," *American Behavioral Scientist* 19, No. 5 (May-June 1976):573.
4. Frankena, *Ethics*, p. 26.
5. Wakin, "The Ethics of Leadership," p. 577.
6. Much of the discussion on this point is drawn from Peter Berger, *Pyramids of Sacrifice: Political Ethics and Social Change* (New York: Anchor Books, 1976), p. 249.
7. Ibid.
8. Frankena, *Ethics*, pp. 24-25.
9. Immanuel Kant, *Foundations of the Metaphysic of Morals*, trans. L. W. Beck (New York: Liberal Arts Press, 1959), p. 19.
10. See Donna B. Ayers and Stephen D. Clement, *A Leadership Model for Organizational Ethics* (Indianapolis: U.S. Army Administration Center, 1978), p. 89.

3 THE NEED FOR MILITARY ETHICS

The need for ethics in any area of human endeavor is self-evident. Without some standards of behavior and judgments of that behavior, peaceful human intercourse becomes impossible. In a general sense, the one element that makes human society possible is the expectation on the part of one's fellow men that individuals will observe certain obligations. In the most rudimentary sense, observing obligations in a willing manner is what we call ethics.

All social actions require some regulation or else the potential for stasis increases. Indeed, the propensity for civil violence is magnified in any society in which men form highly organized groups and have at their disposal enormous private resources that can be put to destructive purposes. While social action requires some regulation, the question remains as to where this regulation originates. Some believe that man is inherently self-regulating and that he can be relied upon by his very nature to pursue that which is good. Historically, there is little evidence proving the truth of this proposition. As Admiral James Stockdale has observed: "Humans seem to have an inborn need to believe that in this universe a natural moral economy prevails by which evil is punished and virtue is rewarded. When it dawns on these trusting souls that no such moral economy is operative in this life some of them come unglued."[1] The point is that while ethics is required in order to make human society possible and while regulation of human action is needed to achieve the "good" society, ethics does not really function apart from society. It is men who evolve notions of what is right or wrong, and it is men who observe their ethical obligations. It is men who engage in ethical action, and it is men who render judgments about the ethical quality of the actions of others. Ethics is, therefore, a creation of men, and the need for ethics seems an absolute requirement for human society to exist.

A Special Need

An admission that society requires ethical standards in order for reasonable social intercourse to occur by no means implies that a separate set of ethics should be applicable to the military profession. It is clear, however, that the military does, indeed, have a need for a special set of ethics. At its most basic, the military is a profession; it proclaims itself to be a profession, its members feel it is a profession, and nonmembers recognize it as a profession. At the very least, a profession requires a special set of obligations and requirements that make membership in one profession different from that in another. It is inconceivable that any profession should not make some statement of how it differs from what other people do. Not only must the profession state what makes it different, but it must also encompass an *ethic of service* as opposed to an *ethic of self-interest*. What distinguishes a profession from mere occupations is its special sense of ethics, which must include the requirement that members observe their obligations, not only in addition to their self-interest but also, in some instances, *instead* of their self-interest. Given life and health risks that soldiers are likely to face, the requirement that one may be obliged to observe obligations even unto death truly constitutes a special and unique sense of ethics, obligation, and responsibility.

All professions must have codes of ethics, for it is in observing such obligations that one becomes a member of the profession. There can be no military profession without a code of ethics that does not state an obligation of service to a larger group, society, or higher cause than the profession itself. In a word, the pursuit of "enlightened self-interest" within a profession can never be truly legitimized in its code of ethics. What makes the military sense of ethics different from the ethics of the society at large is precisely this requirement of service instead of self-interest. As long as the military is to remain a profession, its members and the profession itself must put service to its clients above self-interest.

The military perhaps has a greater need for ethics than any other profession because the military task involves the systematic application of social violence. The consequences of unethical behavior within the military environment are potentially far more devastating than within civilian life. The larger society can tolerate a wider scope of unethical behavior, even among its other professions, largely because the consequences of that behavior are likely to be restricted to a small number of people. The consequences of unethical action in the military, especially on the battlefield, can be catastrophic, for they can immediately affect hundreds or even thousands of human beings. Major R. I. Aitken of the Canadian Staff School expresses this view:

> The consequences of a degenerating ethical climate are bad enough in our current time of peace; they would be disastrous in war. War places men under unparalleled pressure, no matter where in the forces they serve. At all levels tough decisions must be made—decisions that can cost lives. There is no room for anything but an eye toward the common good here. Mutual trust is indispensable if the forces are to operate the way they must. The whole structure of discipline and esprit de corps will disintegrate if officers cannot see past their own wants and aspirations.[2]

The tension between individual self-interest and service to the community within the profession must be resolved by the soldier in favor of the community and its values. To have a corps of officers and men guided only by self-interest is to risk enormous ethical and human desolation on the battlefield, to say nothing of the threat such a profession might pose to a democratic civil order.

The special nature of the military task, the systematic application of violence against other human beings, makes the development of ethical standards for its members even more necessary. In a more generic sense, so many countries now have nuclear weapons that military establishments, if not subject to ethical restraints, could sentence whole societies to death.

Military Effectiveness

Ethics in the military profession is needed not only to compel ethical behavior and avoid moral disaster, but also to enable a military organization to effectively engage in combat.[3] The military's loss of some of its traditional values and their replacement with the values of the economic marketplace can lead to the abandonment of ethical precepts, or at least their distortion, to the point that combat effectiveness itself is affected.

As noted earlier, the American military has incorporated a number of techniques, practices, and values first developed by civilian business corporations. The military's tendency to imitate the business world reached a peak with the introduction of the All-Volunteer Force in 1973. It continues unabated today. In all the arguments for efficiency, cost-effectiveness, and administrative streamlining that have been marshaled to justify the military's adoption of business values and techniques, one fundamental truth has been overlooked: from the point of view of combat effectiveness, the adoption of business practices, values, and ethics has been the equivalent of a military disaster.

For example:

- During the Vietnam War, the military tried to provide all of its officers with the opportunity to command. As a result, officers served only six months in combat while their soldiers served twelve. Morale fell badly, and several hundred officers were assassinated by their own men.
- In an effort to economize on replacement costs, the military adopted the DEROS system—modeled after the system used in the auto industry to replace spare parts—where individuals instead of units were fed into the "replacement stream." Combat units quickly became associates of strangers, and the rate of mutiny and combat refusals rose to the highest point in American history.
- In order to train more officers for staff work, large numbers were sent to staff schools. Coupled with the "up-or-out" system that prohibits an officer from being a career combat commander, the rate of leadership turnover—"personnel turbulence"—increased enormously so that in 1979 fully 80 percent of the military's personnel changed assignments. Leaders became remote from their men, combat unit-effectiveness levels fell, and desertion rates rose.

Many of the same practices developed by business enterprises which work so well in the civilian sector have actually undermined the combat ability of the military. The instances described above are only three in a list that grows longer every day.

The key to any successful military profession is its ability to develop small combat units that will remain intact and perform their mission under the terrifying stress of battle. Unit cohesion is not the result of weaponry or even the quality of troop training. It is the result of strong bonds of shared attachment among members of the battle group, which in turn are a function of sharing the same hardships, the same risks, and a common fate and of having one's leaders in clear evidence during combat. In this basic sociopsychological sense, the motivation of military units has not changed throughout military history.

The management techniques and values adopted erode these conditions by destabilizing leadership elements, disrupting unit stability, or demonstrating that the risks of combat do not fall on leader and led alike. As a result, military units cannot sustain the pressure of combat and crack apart. Neither ideology, home front support, nor the organizational imperatives of the system itself will prevent disintegration. To be sure, small unit disintegration occurred on a large scale in Vietnam, and it will happen again in future conflict if the wholesale acceptance of business practices, values, and ethos is not halted.

The subtle transformation of the military profession and the officer corps began during World War II with the influence of then Chief-of-Staff George C. Marshall. Faced with the necessity of pulling together the multiple centers of economic and social power to fight the conflict on a grand scale, Marshall turned to the only model available to him which had some experience in the field of organization and was consistent with the values of democracy and free enterprise—the embryonic business corporation. General Marshall could hardly have foreseen that his choice of model would, in time, weaken traditional military values.

The curiously symbiotic relationship between the military structure and the modern business corporation which began during World War II has continued to the present. The rise of huge defense budgets needed to support a large standing military establishment has much to do with both how and why this relationship developed and sustains itself. Throughout the 1950s, the military adopted more and more of the internal control, auditing, and evaluation techniques of the business corporation. By 1960, the military had so many civilian business community apparatuses that the appointment of Robert S. McNamara as secretary of defense, a man whose only previous experience had been to produce automobiles, seemed unusual to no one. Himself the very model of the successful executive, McNamara signaled the rise of a new breed of military managers that would now staff the military establishment.

The Need for Military Ethics   61

McNamara is remembered chiefly for his insistence that "good business practices" be applied to the design and purchase of military weapons systems. Yet, the symbolism of McNamara's influence ran much deeper. He was the ideal corporate man, and during his tenure as secretary of defense, the military moved ever closer to the modern business corporation in concept, tone, language, and style. The military officer became thoroughly identified with the corporate executive to the point where the functions and responsibilities of command were perceived to be identical to the functions and responsibilities of management. More and more of the military's officers were sent to graduate schools to receive advanced degrees, the overwhelming majority of whom received degrees in business management or administration. The traditional elements of the military way began to collapse under the impact of new administrative skills, staff reorganizations, computer models of decision-making, and the redefinition of the criteria needed to succeed within the profession itself. Military leadership in the traditional sense became obsolete; indeed, it became unnecessary. The machines, new administrative doctrines, and a core of military managers would show the way. The era of the automated battlefield had arrived.

Had the military been more selective in its adoption of business technology, it is questionable whether serious damage would have been done. But the fact is that it absorbed not only the technology of the business world but also its language, style, and eventually ethics. However gradually and subtly, the military ceased to be a unique element set apart from society. The symptoms of the transformation are everywhere. The extent of the metamorphosis can be seen in the practice of referring to lower ranking officers, those expected to carry the burden of leadership with combat units in the field, as "middle-tier managers." The officer corps has come to believe all too readily that leadership and management are one and the same thing, and that the mastery of the techniques of the latter will suffice to meet the challenges of the former.

Military systems, especially the small unit subsystems that are expected to bear the burden of killing, are categorically unlike anything in the business world. No one truly expects anyone to die for IBM or General Motors, but the expectation that the soldier will carry out his duty even unto death, that he will live up to his "clause of unlimited liability," is very real in the military. Consequently, the forces that compel an officer to fulfill his obligations to himself, his command, his superiors, and his profession are categorically different from those that press the corporate official to fulfill his obligations. Moreover, the circumstances under which the obligations of the soldier must be met are extremely different from those of the corporate executive. The military profession, especially with respect to its combat forces, often has failed to realize that combat leaders are not, in truth, managers of any sort.

Military organizations that are successful in withstanding combat stress require high levels of individual identification with community goals to compel individual action. This belonging and uniqueness define a truly cohesive military unit, and motivate the individual solider to stand and fight and to risk death in the service and protection of his comrades. The adoption of business ethics within the military environment portends disaster, for in a free enterprise democratic system, business "ethics" are dictated largely by cost-effectiveness which, in turn, is directed solely to the maximization of profit. The free enterprise system thus constitutes the negation of ethics in the sense that the pursuit of self-interest at the level of the individual is the highest value. Moreover, such pursuit is expected to result in the emergence of a community of interests and values at the organizational level. *Accordingly, the individual has no direct responsibility for developing and following ethical norms that address only community goals.* The "ethics" of business is not really ethics at all as much as a doctrine of the rapacious pursuit of self-interest. The individual is expected to become predatory man. And, unfortunately, this sense of business "ethics" has all too often served as a model for the military.

With the military's absorption of this perspective, the traditional values of community sacrifice, service, and dedication to one's comrades has surely been undermined. As a consequence, career management becomes the ultimate means to the ultimate value—promotion. The cumulative impact of this change has been the rise of the "officer as entrepreneur"—the man adept at managing his own career by manipulating the system, mastering its technology usually defined in terms of administrative and managerial techniques, having his "ticket" punched, and achieving the "right" assignments in order to qualify for the next promotion. An officer who can master these avenues of career advancement is rewarded with the ultimate goal: promotion to the rank of general.

The entrepreneurial officer and the ethics that motivate him constitute one of the major problems afflicting the military profession today. The adoption of the ethics and practices of the entrepreneur constitutes a severe and corrosive force within the military profession because it eats at the very foundation of the profession, a sense of selfless service. It is encouraged and sustained by a hundred different policies and practices, ranging from an officer evaluation system that almost all agree is inflated and measures nothing except trivialities to the practice of rotating officers through a series of assignments and schools in pursuit of the preposterous doctrine that every subaltern is a potential chief of staff. The insane doctrine of "up-or-out" which forces good officers to leave the service after twenty years of honorable service instead of allowing them to remain for thirty years or to remain at one rank for long periods without being passed over also hurts.

If the ethics of business enterprise is not effective in producing cohesive military units, what is the military profession to do? What guidelines if not cost-effectiveness, efficiency, and good management should the effective military adopt? Of course, some business techniques are useful. In the management of such things as spare parts, munition, and food supplies, modern computerized management techniques work well and ought to be used. But in the area of developing fighting units, good leadership is required, and good leadership is not the same thing as good management. By and large, any managerial technique that erodes the personal ties between soldiers and their leaders—rapid turnover in officer assignments, individual instead of unit replacement, proliferation of staff assignments, use of administrative control devices at the small unit level, and centralization of promotions—should be abandoned.

In general, the military profession needs models of organizational development that stress personal ties, social interaction, ethics, stability of leadership, and community identification with the profession itself. The basic point is that such "premodern" organizational forms rely heavily upon the interpersonal interaction and identification of its members with one another rather than upon an amorphous, faceless bureaucratic structure to compel suitable behavior. To be sure, one cannot dismantle the military bureaucracy, nor would that be advisable. What must be done, however, is to restrict the penetration of those practices and values which might be appropriate to noncombat elements of the profession so that they do not corrode the basis of the fighting units.

Not all of the ills of the military profession can be attributed to a lack of ethics. Nonetheless, many of these ills have been allowed to develop and continue precisely because members of the profession have acquiesced in policies they felt to be wrong. Perhaps they have done so in defense of their careers or, more likely, because there was no organizational support within the profession itself. Ethics in the military profession must operate within a far more complex organizational setting than in other professions, and within that organizational setting rewards and punishments are meted out in support of ethical precepts. Without organizational support for ethical precepts, individuals cannot easily resist the power of the organization. Until the military develops its own code of ethics, the soldier confronted with an ethical dilemma will inevitably find himself arrayed against the organization and its formal values of success, personal advancement, and loyalty to superiors. In an ethical conflict, organizational forces will probably win out, for it is the rare individual who stands up for his ethical beliefs—and such a person may find himself isolated and perhaps even driven from the profession.

The military profession has a poor record of developing and institutionalizing support for ethical dissent to its policies. A series of problems traceable to ethical failures during the Vietnam War continue to plague the military today. During the Vietnam conflict, the general failure of soldiers to object to many operational policies of that war created a series of conditions that engendered severe ethical problems. Some of these problems are still with us.

The Vietnam Example

During the Vietnam War, American military units, especially ground units, revealed unmistakable signs of small unit disintegration. Some of the more obvious signs included high desertion rates, high rates of drug use among the troops, assassination of leadership elements, and high rates of mutiny and combat refusal. With regard to desertions, for example, the Army alone suffered a desertion rate of 38 per 1,000. When measured against pre-Vietnam standards, the desertion rate between 1965 and 1971, the years of the most intense combat, rose by 468 percent.[4]

That the military was failing to deal with its problems is shown in the failure of the new managerial leadership to come to grips with the drug problem among the soldiery. As discussed in Chapter 1, drug use among the troops was extensive, with official data showing that some 28 percent of the troops admitted to regular use of hard drugs. It has been estimated that approximately 600,000 men acquired the drug habit in Vietnam.[5] While the extent of drug addiction remains debatable, the degree of drug use does not.

Perhaps there is no clearer evidence that the officer corps failed to perform effectively during the Vietnam War than is found in the data concerning the assassination of leadership elements. During that war the number of leadership elements, officers and NCOs, that were assassinated by their own troops was the highest in American history. The data reveal that at least 1,016 officers and NCOs were murdered by their own men.[6] This counts only those individuals who were caught and tried. Unofficial conversations with judge advocate general officers suggest that the rate may have been much higher. Given that slightly over 4,000 officers lost their lives during the war, it may be that almost 20 percent of these casualties occurred at the hands of our own men! The number of combat refusals and mutinies was also very high. While hard data are difficult to come by, data provided by the U.S. Senate in official testimony suggest that as many as 254 mutinies and combat refusals may have occurred in the single year 1971.[7]

The high rates of drug use, assassination, desertion, and mutiny during the Vietnam conflict are clear indicators of a serious lack of cohesion among combat units. At the same time, these pathologies continued for a long time with no officers protesting or resigning, or with very little corrective action taken. Thus, there was a clear ethical failure on the part of the profession itself. Apparently, the leadership elements failed to see that their roles involved special calling and a special ethics. Whether or not one can blame the leaders' lack of ethical action for these pathologies, their perpetuation over ten years of warfare can certainly be traced to the officer corps' faulty understanding of its own special role within the profession.

Negative Institutional Factors

The weakening of small unit cohesion within military units during Vietnam was exacerbated by several institutional and systemic practices of the military itself, practices that continue today. These policies often seem designed to deliberately undermine the social and psychological ties that bind men together in combat. As such, the erosion of cohesion and combat effectiveness can be seen not so much as an aberration or failure of the system but as a logical consequence of its normal operations. Most specifically, it can be seen as a logical consequence of the adoption of a series of practices taken directly from the business world and applied to the military.

A clear example of the kinds of systemic practices that tend to erode cohesion and effectiveness is the use of the DEROS system of individual replacement instead of unit rotation. When individual soldiers instead of units are replaced in combat, it becomes almost impossible for soldiers to develop strong ties to one another or to their leaders. At the same time, the DEROS system increases the degree of personnel turbulence, especially for those people who are small unit leaders. For example, the use of six-month tours for company commanders, platoon leaders, and battalion commanders while the troops spent a full twelve months in combat isolated unit leaders from their men as well as conveyed to them a certain sense of basic unfairness as to who was carrying the burden of battle. While tours of duty have been lengthened, the fact remains that the eighteen-month anticipated tour is still too short and personnel turbulence still too high. The atmosphere in many American military units is still that of an association of strangers led by officers of unknown quality.

Another practice that weakens unit cohesion and that continues very much in force is the inflation of the numerical strength of the officer corps. By historical standards, the Army officer corps, bloated to 16 percent of total military strength during Vietnam, was far too large. It is still too large today, accounting for over 11 percent of total strength.[8] Historically, effective military forces have rarely exceeded 4 to 5 percent of total strength for their officers. The American Army exceeded that standard by almost four times during Vietnam and still exceeds it by more than twice as much today. On top of this, far too many staff officers were far too visible to the troops. It will be recalled that Vietnam was a war of base camps where combat forces sallied forth for short periods to engage the enemy and returned to camps in which they continually encountered hives of staff officers engaged in support functions, none of which ever seemed to involve the risk of death. Moreover, in today's profession, the bloated officer corps, like any bloated bureaucracy, represents an institutional condition that serves to diffuse responsibility and lessen ethical accountability rather than focus it.

The institutionalization of the "up-or-out" system of promotion makes it impossible to establish a stable core of combat commanders with long periods of assignment with their units. This inevitably contributes to a lessening of cohesion in battle units. The change in the military from a vocation to that of another occupation has also eroded effectiveness. Once the officer corps began to see itself as managers pursuing careers and to regard war as one more opportunity to gain advancement, ethical problems were certain to arise. The performance during Vietnam showed how thoroughly the penetration of entrepreneurial ethics adopted from the business community had gone. At this point, of course, the erosion of the military as a special calling was nearly complete. Compounding the problem has been the failure to develop and enforce a code of ethics for the profession itself.

Taken together, these systemic practices began to break down unit cohesion and combat effectiveness. The problem of ethical failure within the officer corps was evident enough as it failed to take steps to bring the impact of these negative practices under control. The failure of the officer corps to protest continually failing local policies over ten years of warfare stands as mute testimony to its inability to come to grips with its own ethical responsibilities. General Douglas Kinnard, in his book *The War Managers*, undertook a study of officers, mostly high ranking, who served during Vietnam and were responsible for executing questionable local policies. He reports that almost to a man these officers now have serious doubts, and even ethical misgivings, about the policies they were asked to carry out. Yet, not a single officer in this group either resigned or retired in protest; all carried out policies regardless of their personal feelings about them. Here is a clear indication that the profession, at least the officers, may have lost its ethical compass; the officers did not speak out against those policies they believed not to be in the best interest of their country or in violation of standards of professional ethics. In the end, the Vietnam War demonstrated how massively the careerist and entrepreneurial values had penetrated the military. It also indicated that these new values eroded traditional ethical notions of special trust and confidence. The need to rediscover and reestablish these values is intimately connected with the need for an ethical resurgence within the profession.

After Vietnam and the AVF

It might be argued that the Vietnam War was a unique experience in history, and, thus, the reaction of the military profession to it was idiosyncratic. As a corollary to this notion, it might be said that the difficulties that were encountered as well as the pathologies that emerged were idiosyncratic and represented no discernible trends. But evidence from several analyses suggest that this view is demonstrably false.[9] The conditions of warfare in Vietnam were not unique in any sense, and many of the pathologies that emerged were not idiosyncratic but predictable symptoms of the larger transformation of the military away from a special profession to merely another occupation.[10] When that transformation occurred, no one should be surprised to find middle-tier executives, nay officers, behaving much the same as the self-interested executive in the business community.

An examination of the military since the end of the Vietnam War shows that many of the systemic difficulties which contributed to a lack of military effectiveness remain quite visible today. For example, the DEROS system is still used, and unit rotation is still not practiced. Personnel instability is still too high, with 80 percent of the soldiers having changed station in the single year 1979. At present, the military is attempting to stabilize tours of duty at twenty-two months, which itself is too short, but at least it is recognized that the problem remains unchanged. The "up-or-out" policy is being exacerbated by new forced retirement policies. Under present circumstances of managing the disciplinary rate, the level of discipline among the general soldiery is low, and the military's image as an occupation instead of a profession requiring special sacrifice has increased greatly with the introduction of the All-Volunteer Force.

Almost ten years have passed since the American forces left Vietnam; yet, many of the institutional patterns, practices, policies, values, and behavior associated with that conflict remain essentially unchanged. Moreover, there has been no swell of protest, dissent, or resignation among high-ranking leadership elements in an effort to bring about change. Far too few individuals have protested the continuation of failing policies, or even the initiation of new ones which, in some instances, have proven even more devastating. In this last instance, I refer, of course, to the introduction of the All-Volunteer Force.

In terms of cohesion, discipline, and fighting ability, the AVF has been a great disaster. Even so, the military leadership continues to publicly support the AVF. For instance, from a study of the AVF, it is immediately evident that the military is not carrying out its traditional responsibility for good order and discipline among the soldiery. Drug use remains very high. In addition, over half the troops in a recent Army survey report they use as many drugs on duty as off. Secretary of the Army Clifford Alexander, in testimony before Congress, did not bother to deny high rates of drug use among the soldiery, noting that drug use was not a problem in the military because the figures were about the same as for the society at large! Indeed, it most certainly does reflect a similar drug use rate. But because drug use is tolerated and permissible in the larger society to tolerate drug use in the miltary totally misunderstands the special nature of the military profession as well as the risk associated with drug use on the battlefield. Moreover, the AVF has attracted lower quality raw recruit material from which to fashion good soldiers. During the Vietnam War, only 10 percent of the soldiers were classified in category 3B or below. Today that figure is at least 50 percent and is climbing every year. The secretary of the army's solution is to stop using the classification categories altogether!

With regard to the question of training and trainability, in 1978 the Military Personnel Center (MILPERCEN) estimated that the overall MOS (military occupational specialty) shortfall approached 40 percent while the Beard Report produced by the House Armed Services Committee indicated that it was even higher in critical combat specialties.[11] Poor quality troops are difficult to train, especially with the equipment of the military becoming more sophisticated with each passing day. The Army Readiness Study of 1980 showed that approximately 40 percent of the troops were inadequately trained for the military jobs they would have to carry out in combat. According to a report on skill qualifications released by the Senate Armed Services Committee in 1980, the failure rate for aviation mechanics was 91 percent, food service personnel 75 percent, signalmen 46 percent, field artillerymen 43 percent, missile maintenance men 39 percent, transportation drivers 83 percent, communicators 69 percent, and military intelligence personnel 51 percent.[12] At the same time, since the advent of the AVF desertion rates continue to be a major problem, despite creative bookkeeping, and AWOL rates have increased by approximately 60 percent. Thus, from the perspective of trainability and combat performance, the AVF has been somewhat less than a success. In addition, the rate of honorable discharge is among the lowest in our history, since individuals who in the past would have been charged with criminal offenses are now administratively discharged. Attrition rates for the first year approximate 38 percent with most of the soldiers being allowed to leave for disciplinary or fitness reasons. Perhaps the most basic indicator that the AVF is not working is that the troops themselves do not feel attached to their comrades or units. Moreover, they do not trust their officers or NCOs. In 1979, a CBS poll taken among soldiers of the Berlin Brigade, one of the military's top units, found that 53 percent of the soldiers did not trust their officers or comrades and would not be willing to follow them under combat conditions.[13] In the face of the evidence, then, it is difficult to conclude that the AVF is doing anything well except raising questions as to how to deal with the disasters it has engendered.

While the disastrous conditions of the AVF have been in evidence for some time, no public dissent over the issue has arisen from within the profession. Hence, it may be concluded that the profession is at least unclear as to its ethical obligations to speak the truth to its civilian superiors. This failure to speak out and to point out serious problems again points up the military's loss of the special sense of service. Why the silence? It may well be that the penetration of civilian and entrepreneurial values has been so thorough and so complete that one of the main ethical supports of the military profession, that a soldier speak out when he is witness to conditions dangerous to his command, his men, or his country, has simply been eclipsed by the requirement to become a team player, to "get on board," and to be loyal to one's superiors. The profession's ethical moorings, weakened by the trauma of Vietnam and by the business infiltration, may have finally torn loose. With these conditions, developing and enforcing a precise code of ethics for the profession emerges as an urgent necessity.

The Organizational Malaise

Many of the problems evident within the military today exhibit the classic symptoms of an organization that has transformed itself from a public service profession into a free enterprise bureaucracy. Organizational analysts have developed several indicators that define when a public service profession, in this case the military profession, moves away from its ideal of service and becomes a mere vehicle to advance the careers of its members. An examination of those indicators shows that signs of organizational corruption are evident in the military today.

Those who have served in the military over the last twenty years immediately recognize the characteristics of organizations that have lost their ethical moorings.[14] An institution that has become merely another entrepreneurial bureaucracy is characterized by the following traits: (1) the organization professes an external code of ethics that is contradicted by internal practices; (2) internal practices encourage, abet, and conceal violations of the external code; (3) prospective "whistle-blowers" are intimidated into silence; (4) the few courageous outspoken men have to be protected from organizational retaliation; (5) collective guilt finds expression in the rationalization of internal practices and (6) those whose role it is to reveal corruption rarely act, and when forced to do so by external pressures, they excuse any incident as an isolated rare occurrence. The Army War College Study on military professionalism noted case after case in which all or some of these instances seemed to apply systematically to the military profession. The clear implication of that study is that the military profession has ceased to be a true profession and has already become transformed into one more bureaucracy that serves to advance the self-interest of its members.

To be sure, some of these pathologies are likely to be found in almost any profession at any time, but too many of them are evident within the military profession at any one time. It indicates that the military may have lost sight of its mission and that a well-developed sense of ethics may have also been eroded under the pressures of increased civilianization, the penetration of managerial values, the trauma of Vietnam, and the impact of the All-Volunteer Force.

As observed earlier in this book, the United States military has never had a formalized code of military ethics. Instead, it has relied upon the officer's oath of commissioning, the Code of Conduct, and the informal motto "Duty, Honor, and Country." All three are inadequate to the task of inculcating and reinforcing a special sense of ethical responsibility required of the soldier. In the first instance, the oath of commissioning is sworn to only once in a lifetime and is never renewed. It cannot serve as a continual reminder or even a ceremonial reminder of the special obligations of the military. The Code of Conduct, while expressing noble ideals of military behavior, is relevant only to wartime service after a soldier becomes a prisoner of war. It is curious that we should have a code of conduct for those who become prisoners but not for those who remain at service in the profession. Finally, the motto of "Duty, Honor, and Country" is too vague, and is at best poorly articulated and even more poorly understood by most soldiers to serve as an ethical code. Its application on a day-to-day basis is unclear, as even the staunchest defenders of the motto point out.[15] In the end, it is clear that the military simply lacks many of the ceremonies and institutions needed to sustain a sense of special calling and responsibility that should rest at the base of any code of military ethics.

At the same time, a series of forces have converged within the military's complex bureaucratic organization and eroded its ethical center. Philip Flammer has identified some of the forces that work against sustaining any notion of special military ethics.[16] The first is the courting of power. A profession that has been bombarded by civilian values of careerism and success at any cost tends to place an enormous premium upon promotion and other status rewards. This is especially so when the normal term of service forces those who are not promoted to be identified by their peers as failures. Unlike the Canadian and British systems which allow individuals to remain on active service at a fixed rank for up to thirty years and still retain the respect of their peers, the American "up-or-out" system encourages the courting of power. It forces the soldier to constantly run in the competitive promotional rat race and to judge one's success or failure as a military professional largely in terms of promotion. This insane competition binds members of the profession together on no basis except that of fellow competitors.

Careerism, Flammer's second force, is a natural consequence of the military's shift from professionalism toward an entrepreneurial bureaucracy based in self-interest. All talk of a profession merely becomes a disguise for careerism and signals a change from the emphasis on service toward self-interest, with the focus moving from the responsibilities of trust and duty to rights of position. Careerism is internally generated since the basis of service as being other than self-interest requires a sense of special ethics which is clearly absent.[17]

Other forces include excessive ambition and conflicts of loyalty. As mentioned earlier, in the military, the concept of loyalty is most often understood as loyalty to one's superiors. But what it *should* mean is loyalty to the larger profession, to the ethical precepts that form its center, and most strongly to the Constitution and society which the profession serves. As every Roman soldier knew, one ought never to confuse *fides* ("loyalty") with *obsequium* ("obedience").

Other tendencies within the military threaten its ethical center. One is excessive concern for image in which career success often depends more on one's ability to look perfect rather than to perform well. Camouflaging mistakes tends to become widespread. Within the profession, it has become commonplace to accept the doctrine that it is worse to admit a mistake than to make it. The concern for image leads directly to an unwillingness to admit error and to harden policies into dogma. All of these conditions tend to diminish the element of choice which is so crucial to willingly fulfilling one's obligations. They also tend to reduce ethical action because they substitute compliance to organizational imperatives for ethical judgment as the basis for action within the profession.

These pathologies are endemic to the military bureaucratic system, and they will not easily disappear; they must be directly confronted and dealt with by the membership. If one is to resist the inertia of organizational form, to say nothing of the other forces affecting the profession, the military must create mechanisms for instilling a sense of ethics in its members and for reinforcing it on a far more regular basis than it has done to date. To take the commissioning oath only once in a lifetime is not enough; to rely upon an individual's sense of ethics acquired in the larger society is also not enough. There is a clear need for more systemic action by the profession. The most basic requirement is the creation of an ethical code and giving it a central place. The code must spell out the obligations which all members of the profession must observe if they are to become and remain members of the special brotherhood of arms.

A Cry for Help

As far back as 1971, Major General William Lynn called for the creation of a professional ethics board comprised of retired and serving officers whose first task would be to formulate a code of military ethics and to establish mechanisms for enforcing these standards. The former chief-of-staff of the Army, General Maxwell Taylor, also called for such a code when he said:

After surveying the many facets of this issue I conclude that it is worth the effort to undertake the formulation of an officer's code possibly as a first step toward one of wider scope for the entire military establishment. It would proclaim to the world what the military profession stands for and by what standards it accepts judgment.[18]

Admiral William Mack, former superintendent at the Naval Academy, also called for the establishment of an ethical code upon his retirement, a code that would apply to the entire Navy. The need for a code of ethics was made clear in 1970 when then chief-of-staff of the Army, General William C. Westmoreland, directed the commandant of the Army War College to undertake an analysis of the "moral and professional climate in the United States Army with a particular view toward evolving solutions to correct this problem. . . . I would particularly like developed an officer's code which would serve as a kind of guide for all officers in exercising their authority and performing their duties."[19] The result of this directive was the Army War College Study of military professionalism released in 1971.

The Army War College Study is too well known among students of the military profession to require detailed recounting here. The findings relative to the subject of ethics can be summarized briefly as follows:

There are widespread and often significant differences between the ideal ethical, moral and professional standards. . . and the prevailing standards.[20]

* * * *

The variances between the ideal standards and the actual or operative standards are perceived with striking similarity by the cross-section of the officers queried during the conduct of this study.[21]

* * * *

The officers queried were concerned about the unethical practices they observed.[22]

* * * *

Variances between ideal and actual standards are condoned if not engendered by certain Army practices.[23]

The study shows that the military profession's inability to define its ethical center has led to widespread unethical practices. These practices have created a wide gulf between the presumed ethics of the profession and the actual behavior of its membership. The study clearly identifies the problem when it notes that:

The Need for Military Ethics   75

The most frequently recurring specifying themes describing the variance between ideal and actual standards of behavior in the officer corps include selfish promotion-oriented behavior, inadequate communication between junior and senior, distorted or dishonest reporting of status statistics or officer efficiency reports; technical or managerial incompetence; disloyalty to subordinates; and senior officers setting poor standards of ethical and professional behavior.[24]

Perhaps the study's most interesting conclusion, considering it was undertaken immediately after Vietnam, was that there was no evidence to suggest that "contemporary sociological pressures" such as the antiwar movement or the failure of the war in Vietnam were the primary causes of the ethical problems that surfaced during the study.[25] Indeed, the study could find virtually no *external* reasons for the unethical behavior. It concludes that "the problems are generated internally within the Army itself and will only be solved as we deal with those problems honestly and directly."[26]

The War College Study demonstrated a clear need for a code of military ethics to guide the profession as well as a recognition of this need by the officer corps itself and, indeed, by all ranks. The study stated that the profession should attempt to establish a specific code of military ethics that would, as General Taylor suggested, constitute those standards by which the profession would accept judgment of its actions.

In addition to the Army War College, the Air Force undertook a study on military ethics. The Squadron Officers Study of 1978 revealed the same conclusions among Air Force officers, namely, a need and desire for a professional code of ethics to guide professional behavior.[27] While the Navy has yet to undertake a similar study, the concerns of Vice-Admiral Mack cited earlier indicate the presence of the same problems in that service as well. Admiral Mack's comments were more recently echoed by Admiral Stockdale, the former president of the Naval War College, in several articles calling for a renewed emphasis on ethics for the profession.[28] All of these studies affirm that major difficulties with the ethical climate of the profession continue and that to correct them would require an unambiguous standard of ethical behavior.

A study conducted at the Army War College by Lieutenant Colonel Melvile A. Drisko in 1977[29] is worth exploring in some detail because it replicated the findings of the earlier War College Study. The Drisko study, although essentially concerned with the Army, documents the ethical problems that exist throughout the armed forces. Indeed, the study amounts to no less than a cry for help from within the profession itself for the establishment and enforcement of ethical standards before it is too late.

The study notes that the profession has had at best a spotty approach to the subject of military ethics. Courses in military ethics are rare in major military colleges and staff schools. In most instances, they are not required and most often constitute only a small discussion topic within a larger course context. Within the curriculum of the Army Command and General Staff College, the entire subject of military ethics is included in a two-day subcourse which, students point out, all too frequently degenerates into a discussion of "war stories" instead of specific instruction in ethics. Of course, since the profession has no clearly established code of ethics or suitable instructors, it is hardly surprising that there are few courses. Drisko is correct when he notes that "for the most part ethics doesn't enjoy the same priority or emphasis as a dozen other military subjects."[30] This is a major shortcoming in view of the connection between the profession's sense of ethics and its ability to carry out military operations.

The Drisko study interviewed 2,215 officers of all ranks at ten different military installations in the United States. Its findings pinpoint the profession's ethical weaknesses and can be applied to the entire profession, regardless of branch of service. Drisko found that:

- 72 percent of the officers thought "the subject of ethical behavior is an important issue for officers today."
- 78 percent felt that "the military took inappropriate action against officers who act unethically."
- 65 percent of the officers felt that current programs about ethics in the military school system were "moderately to very ineffective." 72 percent of USMA graduates felt the same way.
- 77 percent of the officers felt that "more to much more" emphasis should be placed upon ethics in the service school system than is now the case.
- 66 percent felt that unit training programs in ethics were "moderately ineffective to completely non-existent." 80 percent of USMA graduates agreed.
- Only 37 percent of the officers thought that "Duty, Honor, Country" was "moderately to very effective" in promoting ethical behavior. 47 percent felt it was "moderately to very ineffective."

- 55 percent agreed that the military needed a formalized code of professional ethics.

Taken together, the findings strongly reaffirm those of the earlier War College Study and the Air Force Study. The officer corps itself feels that something is seriously wrong with the ethical climate of the profession, that change is needed, and that the best way to begin this change is to formulate and then enforce a code of professional ethics. In fact, all of the studies done by the military over the last ten years document the present system as inadequate to deal with ethical problems and suggest that there is a need for a code of professional ethics, for formal training in ethics and moral reasoning, and for clear moral examples from superior officers. As it prepares to meet the challenges of the future, the profession realizes it must take another ethical direction if it is to serve the country well in the years ahead.

Conclusions

No structured human society is possible without some regulation or limit. From this perspective, ethics must apply to all organized social structures, including the military. As a professional group, the military is by definition an institution of public service and not one of self-interest. Every institution of public service requires a code of ethics which delineates the specific obligations that transcend the individual's self-interest. If the military is to remain a profession, it is imperative that it formalize a code of professional ethics. In addition, it needs a code of ethics because the consequences of unethical behavior within the profession of arms, in this nuclear age, are potentially fatal to whole societies.

When the managerial and entrepreneurial values of profit and self-interest replaced notions of sacrifice and public service in the armed services, major problems were created which were only exacerbated when the AVF was established. If one analyzes those aspects of institutional behavior indicating the transition of a public service profession away from its proper role toward one of self-interest, many of these indicators of organizational corruption are evident within the military profession.

Finally, the officer corps itself has called unambiguously for an ethical renaissance. If the military profession is to retain its professional status the need for an ethical code is desperately obvious. Even if we disregard all other reasons for such a code it remains that when those who have served within the profession, those who are closest to it, are prepared to examine their own professional lives, find it wanting, and openly call for a code of professional ethics to help guide their future actions. This is a cry that we dare not ignore except at grave risk. Indeed, it is a risk that America cannot afford to take.

Notes

1. James B. Stockdale, "Taking Stock," *Naval War College Review* (February 1979): 2.
2. R. I. Aitken, "The Canadian Officer Corps: The Ethical Aspects of Professionalism," *Canadian Forces Staff School* (unpublished paper April 1979), p. 16.
3. Army War College, *Study on Military Professionalism* (Carlisle Barracks, Penn., 1970), p. 33.
4. Richard A. Gabriel and Paul L. Savage, *Crisis in Command: Mismanagement in the Army* (New York: Hill & Wang, 1978), p. 42.
5. Ibid., p. 49.
6. Ibid. pp. 42-43.
7. Ibid., p. 45.
8. Richard A. Gabriel and Paul L. Savage, "Turning Away from Managerialism: The Environment of Military Leadership," *Military Review* (July 1980): 57.
9. One of the best of these analyses is William L. Hauser, *America's Army in Crisis* (Baltimore: Johns Hopkins University Press, 1973).
10. Richard A. Gabriel, "What the Army Learned from Business," *The New York Times* (April 15, 1979), p. 54.
11. *The Beard Report*, U.S. House of Representatives, May 1979, p. 114.
12. *Boston Globe*, September 9, 1980, p. 2.
13. Richard A. Gabriel, "The Slow Dying of the American Army," *Gallery* (June 1979): 96.
14. See Gerald E. Caiden and Naomi J. Caiden, "Administrative Corruption," *Public Administration Review* 37 (May-June 1977): 306-307.
15. Melvile A. Drisko, Jr., "An Analysis of Professional Military Ethics" (U.S. Army War College, Carlisle Barracks, Penn., 1977), p. 20.
16. Philip M. Flammer, "Conflicting Loyalties and the American Military Ethic," *American Behavioral Scientist* 19 (May-June 1976): 589-603.
17. Army War College, *Study on Military Professionalism*, p. 30.
18. Maxwell D. Taylor, "A Professional Ethic," *Army* (May 1978): 21.
19. Army War College, *Study on Military Professionalism*, p. 53.
20. Ibid., p. 30.
21. Ibid.

22. Ibid.
23. Ibid., p. 32.
24. Ibid., p. 31.
25. Ibid., p. 27.
26. Ibid., p. 32.
27. *Squadron Officers Study* (U.S. Air Force: Maxwell Air Force Base, 1978).
28. See the following articles by Admiral Stockdale dealing with the subject of ethics. "Taking Stock," *U.S. Naval War College Review* (February 1979); "The World of Epictetus," *Atlantic Monthly* (February 1978); and "Freedom: Our Most Precious National Treasure," *Parade Magazine* (June 29, 1980).
29. Drisko, *Analysis of Professional Military Ethics*.
30. Ibid., p. 3.

4 PROFESSIONALISM AND THE BROTHERHOOD OF ARMS

In contrast to less developed societies, modern, developed economies have a high degree of role-specification. Thus, men in modern societies generally find themselves in a series of occupational and sociological slots in which certain obligations and behavior are expected of them simply because they occupy these roles. Interestingly, few social roles in earlier societies survived the transition to modern society relatively unchanged. As societies develop, some roles die and new ones are created. Some roles die because they are no longer functional, while others are called into existence by the force of technology. Most fundamental to the creation and destruction of roles in any social structure is the change in values that underlie them. Technology and other forces are major determinants of social roles; yet, none can exist for long if the values that underlie them are eroded. One of the roles that has remained relatively constant over at least two thousand years of social history is that of the soldier.

The Role of the Soldier

Organized society has always had need for the soldier, and the action required of the soldier in order for him to be effective in his role has remained relatively constant over time.

The concept of an officer and a gentleman cannot be dismissed as being an anachronism in today's "me" society. The tasks and obligations of the officer corps have not changed appreciably over the years; indeed in light of the weapons of mass destruction now at our disposal, the military has more of a social responsibility than ever before. Therefore, regardless of the erosion of moral and ethical standards in the world at large, the officer corps must, as a condition of its survival, demonstrate an ethical stance which is above reproach.[1]

While Aitken does not specifically detail how certain characteristics of the soldier have remained constant, there is little doubt that the expectations of the military officer in today's modern society are fundamentally what they were even two thousand years ago.

Whereas the tools and the technology of the military profession have changed, the modern officer must still demonstrate the same kind of personal leadership and character traits, take essentially the same kinds of actions on the battlefield, and establish the same close bonds with his men as did the officers of Rome. The literature on military effectiveness again and again confirms the truth of this statement. In *Face of Battle*, a study of three different battles—Agincourt, the Somme, and Waterloo—John Keegan found that, despite changes in military technology and cultural values, the requirements of military effectiveness, as defined in terms of those things an officer had to do to establish unit cohesion, remained constant.[2] Alan Lloyd in his study of British troops in World War I demonstrated that leadership effectiveness depended on the ability to establish strong personal bonds with one's men and had very much to do with personal bravery, hardship, and sharing risks.[3] The consistency of an officer's behavior as a force for military effectiveness is perhaps best documented in the Shills and Janowitz study of the German Army in 1948.[4] Shills and Janowitz discovered that the battlefield cohesion of the German Army in World War II largely reflected the strong interpersonal relations that developed among men in combat units. Samuel Stouffer, in his study of the American Army in World Way II, came to the same conclusions as did Paul Savage and I in our study of the Vietnam War.[5]

The role of the officer and of what might be called military virtue, and their contribution to military effectiveness, seem almost constants throughout military history. In this sense, the profession of arms even in our modern society is unique in that it represents one of the few social roles that has survived intact. It might even be argued that to the extent that unit cohesion is crucial to the ability of a unit to perform effectively and to the extent that professional behavior contributes to cohesion, then the battlefields of the future, replete with sophisticated technology and lethal weaponry, will reveal an intensity of combat so great that even greater cohesion within units will be needed. The future, therefore, seems to call for an officer corps that is sure of its role and of its actions, and able to apply the lessons it has learned over two thousand years of experience in building unit cohesion. Indeed, in order to secure success in the future, the military profession will have to reaffirm the lessons of the past.

Professionalism

Many of the difficulties found within the officer corps today can be traced to their failure to recognize that what they do is anthropologically different from other tasks and occupations found in the larger society. This confusion of roles and objectives has led many members of the military to pattern their own values and behavior after those of the larger society, and even after specific business enterprises within it. The results have been disastrous.

If one is to understand why the military is different, one must first understand why members of professions incur different kinds of obligations from those of other social organizations that do not enjoy professional status. How is a profession different from other social groups in a society?

Arthur Dyck has done an interesting study on professionalism in which he observes that the very word "profession" is an honorific term denoting that there is something special about a profession that is not shared by other occupations.[6] A profession has certain distinguishing characteristics that set it apart. The difference between a profession and an occupation is that the profession performs a service which is recognized as more important to the society than other tasks generally are. Thus, the medical profession, which deals in the treatment and saving of human lives, is considered to be far more valuable than the marketing of vacuum cleaners. The *service* that the professional renders to his "clients" in an altruistic manner is "the first ethical imperative" of a profession.[7] From the perspective of the military, the fact that it stands as the line of defense in protecting the quality of life in society by its willingness to risk its life and to engage in the very nasty (but necessary) business of killing other human beings suggests that the military meets the requirement of special service. The military profession unlike other occupations performs a vital service to the society as a whole. Indeed, it can be argued that unless the soldiery is strong and the social order protected from attack no other organized social interaction can be secure for long. Whereas American society is based fundamentally on the pursuit of individual self-interest, the American soldier is obligated to carry out his task even to the detriment of his self-interest for the good of the larger society or the profession itself. As Lewis Sorley has cogently pointed out, "in military service involving combat the obligation to serve the general interest is typically to do so *instead* of rather than *in addition to* one's own self-interest, even to the extent of sacrificing one's life."[8] Of course, other professions do not involve the sacrifice of individual self-interest in pursuit of the common interest. Examples are medicine and law which are richly remunerative for their members. In a true profession, individuals must at times carry out their obligations, not only in addition to their self-interest, but instead of it. And this, of course, places at the center of any sense of military professionalism the obligation of sacrifice.

What sets one profession apart from other social institutions is a code of ethics to guide the membership in rendering their service to their clients. Without a code of ethics, there is no standard against which the action of the profession can be measured, and there is no clear statement of what the profession does that separates it from other occupations in the society. Finally, without a sense of what the profession is all about, it is extremely difficult to socialize new members to it in terms of the special obligations required of them. Thus, a code of ethics for the military profession is a necessity and not simply a nicety. For the military to claim that it is a profession and not profess a code of military ethics is a contradiction. Simply phrased, a profession that has no code of ethics to specify obligations and values for its membership is not a true profession. As long as the military does not have a code, it, too, cannot consider itself a true profession. It lacks a critical means for linking the individual behavior of its membership to the expected behavior of the profession in terms of its wider obligations of service.

All professions require the possession of special knowledge or expertise, and all professions lay claim to some degree of special competence. But the achievement of expertise in itself does not constitute a profession. An auto mechanic, for instance, possesses a special competence, and in a general sense he even "serves" the public; but no one would seriously claim that auto mechanics constitutes a profession. Rather, to suggest that a profession requires specialized knowledge is only to point out that its membership must be competent in exercising a special task. Expertise per se does not constitute a truly special quality since in complex modern societies specialization of one kind or another sets members of one task group apart from another as a matter of course. This task isolation is not the equivalent of expertise in the military.

The military profession has requirements for special knowledge and expertise that rests in the very clear function of the military to kill—to carry out the systematic application of violence in the service of the state. This is what military organizations do, and the acquisition of special expertise and competence is linked to those techniques involved in the performance of combat. The techniques of the military profession are not widely available and, in point of fact, can only be legitimately acquired and practiced within the confines of the profession itself. The military certainly meets the requirement that a profession have specialized knowledge and that it exercise this expertise against the background of community service.

Professions also require that performance be subject to rational analysis and to standards of competence as enforced by peers.[9] The military profession is not a priesthood in that its claim to expertise is overtly measurable. Bad armies are defeated on the battlefield; God only knows what happens to bad priests! The military must be competent in the application of its special knowledge, and that competence must be subject to rational analysis and measurement. Performance and standards of competence must be subjected to review by superiors and peers. The profession has a responsibility to police itself, ensuring that its standards of competence are being met and that it can carry out its obligations to the larger community. By implication, those who do not meet standards of professional competence must be driven from the profession. No conspiracy of silence, which is so common in the medical profession, is ever acceptable for the military.

Unlike other professions where there is usually daily opportunity to practice one's skills, the military employs its skills in earnest only rarely, in times of war. In order to maintain professional standards between periods of war, levels of competence must be subject to analysis and judgment by members of the profession itself.

The above analysis implies yet another aspect of military professionalism. Members of a profession must have a certain degree of autonomy so that it becomes self-regulating. Of course, the military can never be self-regulating to the same degree as the medical or law professions and for obvious reasons. The military is an instrument of the state and, therefore, the political masters of the state necessarily and rightly have an obligation to oversee and control military action. Nonetheless, except in those areas of grand strategy which are dictated by the military's political masters, the profession retains an autonomy in which appropriate authorities within the profession itself have direct responsibility for ensuring that it is meeting its standards of competence and ethics. In this sense, the profession is self-regulating. At the same time the military has autonomy of action insofar as it has its own law, its own set of courts, its own mechanisms of enforcement, its own jails, its own food supplies, and so on. It does not rely on the larger society for its day-to-day operations nearly as closely as do other professions.

According to Samuel Huntington, a profession must have a sense of corporateness.[10] Members of the military must believe that what they do is more important than what other professions do—that they are crucial not only to the survival of the society, but also to the preservation of the quality of life. This sense of special corporateness seems to follow fairly regularly if the other elements of professionalism are institutionalized and observed. The sense of corporateness should never be allowed to degenerate into a narrow specificity of outlook. In order to prevent this isolation of the intellect, the military must provide broad training and education which helps the members fit the role of their profession into the larger concerns of the human being. While the soldier must be a proficient technician, he can never be a mere technocrat in the application of his skills. Rather, he must strive to be a humanist.

Members of the profession must understand that to be a good soldier is not the equivalent of being a good human being. They must realize that the requirements of the profession will conflict with other obligations acquired in other aspects of their lives. The recognition of these conflicting obligatons and their solution constitute the nub of ethical action. A profession must never become a refuge for those who abandon their humanity, eliminating all other social or moral obligations in deference to purely professional standards. To carry out obligations to the profession without regard for other obligations is not only to reduce the soldier to an automaton but, also, in some instances, to court moral disaster. A member of a profession is not merely a technician; a true member of the military profession must also be a humanist. Throughout history, one can point to individuals, often general officers, like Von Moltke or Patton who were not only military men but also poets, authors, philosophers, musicians, and athletes. Those who would claim membership within the military profession must understand that the professional experience is not the sum total of either their ethical existence or their obligations.

The Military Is Different

The military profession is distinguished from professions in civil society by four characteristics: *scope of service, level of responsibility, extent of personal liability,* and *monopoly of practice.* With regard to scope of service, the responsibility of the miliary profession is simply greater than that of any other profession. Unlike other professions, the military is responsible for the very survival of the state and its society. Not even the medical profession can in any meaningful sense be construed as being responsible for the very survival of society. Lewis Sorley has noted that "the point is in itself really a very simple one: nations are critically dependent upon their armed forces for survival, and thus the competence of those forces is of greater concern and more general impact than that of any other profession."[11] The military's scope of service, the sheer number of people that the profession must address in its service, is larger than that of any other profession; it encompasses the entire society.

With regard to level of responsibility, no other profession in society has responsbility for the lives and death of such large groups of people. The argument does not speak to the life and death of one's enemies as much as to those of the charges—one's fellow citizens—placed in the care of the military. Moreover, no profession has the awesome responsibility of legitimately spending the lives of others in order to render its service. Sometimes it is easier to meet one's death than to be responsible as an officer for sending men to their death in pursuit of military objectives. When discussing the military profession's level of responsibility for life and death, the tendency is to regard death on the battlefield as somehow accidental. The truth is that the responsibility of military leadership includes the ability to sacrifice human lives in pursuit of battle objectives. The actual trading of human lives for a piece of terrain, a railhead, a bridge, a crossroad, or any other military objective imposes on those men who must make such decisions—and indeed they are made by all members of the profession at all rank levels—an awesome responsibility. No matter how difficult medical or legal decisions may be for members of these professions, it is impossible to imagine a case where a lawyer must be willing to go to jail himself or send other members of his profession to jail in the service of his client, and few medical decisions involve the sacrifice of the lives of other doctors in order to save the patient. In the military, then, the level of responsibility which its members assume at both the organizational and personal level is ordinarily higher (and more frequent) than found in other professions.

For members of the military, the requirement of service is total. It encompasses what Sir John Hackett has called the "clause of unlimited liability,"[12] which involves obliging members of the profession of arms to give their lives in the performance of their professional duties in service to the state. As Saint Paul noted, the military man must be prepared to sacrifice his life and render his service "even unto death." General Sir John Hackett, in *The Profession of Arms*, summed up the extreme degree of personal liability expected of members of the military profession:

The essential basis of military life is the ordered application of force under an unlimited liability. It is the unlimited liability which sets the man who embraces this life somewhat apart. He will be (or should be) always a citizen. So long as he serves he will never be a civilian.[13]

No civilian profession requires the sacrifice of one's life in its service, whereas the military regularly requires it. The clause of unlimited liability separates members of the profession of arms from *all* other professions in civilian life.

Yet another characteristic of the military profession which separates it from others is the military's monopoly on the skills it practices. In most other professions, one may leave one's place of practice and still practice elsewhere. Thus, a doctor may leave one hospital for another or a lawyer one firm for another, or both may practice their respective skills in private practice. But once a soldier leaves the military he is outside the brotherhood forever. Technically, one could become a soldier of fortune, but still one would no longer be part of a profession per se. No other profession maintains a monopoly of practice in the sense that one must either belong totally to it or not practice one's skills at all. While other professions sometimes exclude members from practicing, that is clearly not the same thing as having a member in good standing, whose skills are up to date and who wishes to practice, not being able to practice because of the monopoly on the service offered.

When examined from the perspective of its scope of service, level of responsibility, personal liability, and monopoly of service, it is clear that the military is not just a profession but a profession unique among all others. It is a profession that demands a high level of competence, expertise, and service; perhaps more than any civilian profession. It is a profession that will require, when circumstances warrant, that one be prepared to sacrifice one's life for it. For at least these reasons, the profession of arms is unique. Because it is unique, because it imposes special obligations, and because it requires special men to fulfill them, the military profession must be separate even from the society it serves.

A World Apart

The American military profession has three clients: the Constitution which it has sworn to uphold and preserve, the larger social order it acts to protect, and the members of the profession itself who have assumed the obligation of unlimited liability. When one speaks about military professionalism, one refers to the application of special skills in the service of its clients. There can be no question of a military profession which becomes so self-serving that it does not serve its clients. A profession that serves only itself and not its external clients is no longer a profession but a private enterprise whose concern is the interests of its membership rather than its clients.

To say that the military must serve the larger society is not the same thing as saying that the profession must be completely *of* that society. Quite the contrary. Given the special nature of the military profession and of its obligations and responsibilities, it is clear that a whole range of habits and values cannot be tolerated in the military either because they do not work or because they damage the nature of the profession itself. Thus, *the protection of the society by the profession does not require that the profession become like the society*. And since the society cannot (and ought not) become like the profession in terms of the special obligations it must observe, some degree of separation of the profession from the larger society is required for it to remain a profession and to continue to serve the community.

Professionalism and the Brotherhood of Arms 89

The military must be separate from civil society because there is a range of individual behavior which is accepted, if not approved of, in the larger civil society that is disfunctional to the effectiveness of the military profession. Drug use is an example: it is all too evident in both the larger society and the military. Drug use may sometimes be tolerable in the larger social order, but as noted earlier, within military units it can alter a unit's ability to perform under fire, to generate the bonds that hold the larger unit together, and to operate equipment and exercise judgment under stress. Unless one is prepared to argue that it does not matter whether the military profession is effective in the application of its special task, some forms of behavior must be limited or outlawed.

Another reason why the military must be separate from the larger society is that some of the larger society's values are simply unworkable within the military environment. Some such values take away from the very notion of professionalism and special obligation as they involve ideals of service to the society. Lieutenant General Robert Gard has described this problem:

Vital to combat operations and therefore a necessary part of traditional military professionalism is a set of values which are to some extent contrary to those held by liberal civilian society. Military organization is hierarchical, not egalitarian, and is oriented to the group rather than to the individual; it stresses discipline and obedience, not freedom of expression; it depends on confidence and trust, not *caveat emptor*. It requires immediate decision and prompt action, not thorough analysis and extensive debate; it relies on training, simplification and predictable behavior, not education, sophistication and empiricism. It offers austerity, not material comforts.[14]

Here General Gard is emphasizing a singular fact of the military profession: because it aims at special service to the community at large as opposed to self-interest, inevitably certain basic values in the larger society run contrary to this ideal of service and, therefore, tend to dilute the professionalism of the military way. This problem of the lack of value congruence is particularly acute in Western capitalist democracies and perhaps even more so in American society. Western capitalist democracies hold that the highest value is to be placed on the individual, while the highest behavioral value enshrines the pursuit of self-interest. The pursuit of individual self-interest implies that the individual takes precedence over the pursuit of community interest and, perhaps, even the community itself. These premises are contrary to the military's ideal of selfless service.

If some behavior is antithetical to military effectiveness and if some values are unacceptable because they strike at the foundations of the profession, the profession must protect its members from such behavior and values. The profession must establish and enforce values and standards of action that can serve as ethical reference points for its members. Responsibility for maintaining the military's professionalism rests with the profession itself and its membership. The military must be the keeper of its own flame or be prepared to watch that flame extinguished in the wind of hostile values that blows through the larger society.

Some object, however, that the military's values and behavior must be congruent with those of the larger society. This view is likely predicated more on reasons of ideology and, perhaps, even on a fear of creating a "state within a state" than it is upon solid reasoning. To suggest that the military profession need not be separate from the larger society is to misunderstand the nature of group formation in the social arena. No societal group ever demonstates a one-to-one relationship with the values and behavior of its host society. This is true of football teams, business corporations, fraternal organizations, and other professions, and it is true of the military. Furthermore, there is no necessity for such a complete value congruence to exist. One might even argue that it is quite impossible to establish a direct congruence of values between the larger society and any subsocietal group since a whole range of specific activities undertaken by group members are not at all relevant to any values of the larger society. For example, the decision of a football team to execute a certain play has no bearing at all on the larger values of society. The same may be said of a lawyer and his client. In a word, the value connections between social groups and the larger society are not equally important, nor, indeed, are they equally evident or relevant to the group's behavior in any given instance.

Membership in any societal group also implies a range of behavior that is irrelevant to larger social values. Much of what men do in groups on a day-to-day basis has absolutely no relationship to larger social values, and, indeed, the question of such a relationship never realistically arises. Within the military, for instance, the day-to-day training conducted by a military unit or the requirement that soldiers care for their equipment simply has little to do with larger social values. Within the military, questions of conflicting values do not seem to arise any more frequently than in any other group. To expect all subsocietal groups to maintain values congruent with all larger social values is to misunderstand not only the nature of society and men in groups but the nature of democracy as well. A fundamental distinction between democratic and totalitarian states is that in the democratic there are whole areas of social intercourse in which the writ of the state does not run. Only in totalitarian societies is it deemed necessary for larger social values to penetrate and apply to all aspects of group activity. The major premise of these social orders is precisely the affirmation that certain values explain the behavior not only of the individual in the state but also individuals in all societal groups within the state. Democracy, on the other hand, affirms that there are large areas of human endeavor within the society—indeed even within groups created by the state itself—in which the writ of the state runs in only the most general sense and often not at all in terms of specific day-to-day behavior required of group members. The fear that subsocietal groups such as the military may develop values separate from and even at odds with those of the larger society is rooted in ignorance of the nature of societal group membership and of what motivates men in groups to begin with.

General Walter Kerwin, former vice-chief of staff of the Army, has summed up the dilemma of those who fear that the military will come to constitute a subsociety within the larger society. He states that the military is of necessity different from the civil society. Those who do not understand or recognize this fact miss the point of military professionalism.

We face a dilemma that armies have always faced within a democratic society. The values necessary to defend that society are often at odds with the values of the society itself. To be an effective servant of the people the army must concentrate, not on the values of our liberal society, but on the hard values of the battlefield... We must recognize that this military community differs from the civilian community from which it springs. The civilian community exists to promote the quality of life; the military community exists to fight and, if need be, to die in defense of that quality of life. We must not apologize for these differences. The American people... are served by soldiers disciplined to obey the orders of their leaders, and hardened and conditioned to survive the rigors of the battlefield. We do neither our soldiers nor the American people any favors if we ignore these realities.[15]

The point that General Kerwin is making is a simple extension of the definition of professionalism. Professions require different obligations and responsibilities from their members than one finds in the society at large. Some of these obligations require that certain values, rights, and privileges of the civil society not be allowed free reign within the profession itself. The choice is a simple one: either to separate the profession from the larger society on the grounds that it must remain a profession with special obligations, or to dilute that sense of professionalism and turn the military into one more civilian occupation totally congruent with larger social values, values based on the pursuit of self-interest and antithetical to notions of service and self-sacrifice. Events may force us to choose the second course, but we cannot do so and at the same time maintain our professionalism.

The nub of the argument for separating the military profession from the larger society is reflected in a statement made by President Nixon in 1969. He said, "I believe that every man in uniform is a citizen first and a serviceman second, and we must resist any attempt to isolate or separate the defenders from the defended."[16] Mr. Nixon may have been suggesting that the military cannot be separated from society without it becoming a threat to the state. There is no doubt that the creation of a "state within a state" would be unacceptable, but what is often overlooked is that *it would be unacceptable from the perspective of the professional military as well*. If it is recalled that the nature of military professionalism is to render selfless service to its clients, any military establishment that did not serve those clients would instead become a gang of armed thugs. There is no inherent connection between the evolution of military professionalism based on a degree of necessary isolation from the larger society and the tendency of that profession to become an overt threat to the state. The military profession can never be isolated in terms of its loyalty to the democracy it serves or to the processes that sustain it or in terms of the authority of its elected leaders. The degree of necessary separation concerns only sets of subsocietal values and norms of behavior that are openly antithetical to the preservation of military professionalism and the conduct of effective military operations. In no sense is this isolation extended to the very clients of the military profession itself.

The development of the American military profession has been inextricably bound to the laws, codes, and values of the larger political system. The military was created by an act of law in the name of the larger democratic community. This doctrine was spelled out by the United States Supreme Court in 1866:

> There is no law for the government of the citizens, the armies or the navy of the United States, within American jurisdicton, which is not contained in or derived from the Constitution. And wherever our army or navy may go beyond our territorial limits, neither can go beyond the authority of the President or the legislation of the Congress.[17]

There is no question of the formation of a "state within a state" or a military profession whose origins and legitimate authority do not flow directly from the clients it serves, namely, the Constitution and the community.

The courts have also recognized that, while the military is tied to the other institutions of the democratic civil order, the special nature of the military task requires that certain practices, norms, and values be separate from that order. Justice William Rehnquist issued one of the most recent of a long list of precedents in this regard:

> This Court has long recognized that the military is, by necessity, a specialized society separate from the civilian society.... While the members of the military are not excluded from the protection granted by the First Amendment, the different character of the military community and the military mission requires a different applicaton of those protections. The fundamental necessity for obedience, and the consequent necessity for imposition of discipline, may render permissible within the military that which would be constitutionally impermissible outside it.[18]

Thus, the Supreme Court in a number of decisions has recognized that the profession of arms is different and that it requires different rules for its members.

The issue of the military becoming a threat to the civil order has been grossly exaggerated, often by academics. The question is not one of praetorianism as much as one of harsh pragmatism, of the necessity to make choices enforced by the nature of the military task. Those in our courts and the political system realize that even the most liberal democracies could not long survive without an effective military institution. It is widely agreed that an effective military institution requires a certain separation of values, norms, behavior, and practices which, although commonplace within the larger society, may hamper the successful conduct of the profession of arms.

The need to separate the military from the civilian society has also been recognized in the very organizational structure of the military itself, a structure created by the democratic institutions it serves. What is true of the American military is true of most military organizations, namely, a developed sense of uniqueness not shared by other professions in a democracy, coupled with the realization that this autonomy flows from the special nature of its task. Thus, the American military has its own court system, its own trial procedures, its own law as codified in the Uniform Code of Military Justice, its own judges, its own court of appeals, and even its own prisons and police. While all of these institutions are based on the law of the larger society, the fact that they exist apart from that society, with society's approval, testifies to the historical need to separate the military from the civilian society. General Kerwin was quite correct when he said that "the civilian community exists to promote the quality of life; the military community exists to fight and, if need be, to die in defense of that quality of life.[19] To insist on a congruence of all but the most basic values runs the risk of either militarizing the state so as to provide an effective military force, or civilianizing the military so as to make it as much a mirror of the larger social order as possible. Neither path serves either the military profession or the democratic society it serves.

Challenges to Professionalism

In most Western democracies, the proposition that the military as a profession must be somewhat separate from the larger society it serves has been under assault. The tendency has been to reduce the military's sense of professionalism and to substitute for it the idea that military service is essentially no different than working at any other occupation. At least five challenges are threatening to weaken military professionalism: occupationalism, managerialism, confusion with bureaucracy, specialization, and competing ethical systems.

OCCUPATIONALISM

Occupationalism tends to destroy professionalism by portraying military service as equivalent to working in any other occupation in the civilian sector. Support for occupationalism has come from outside the military and, surprisingly, from within it as well. As Western societies moved towards the post-industrial age, they tended to adopt highly bureaucratic forms of organization. At the same time, they developed highly econometric forms of analyzing the performance of these organizations. As a consequence, some suggest that the military can be treated exactly as any other occupation in terms of analyzing and predicting effectiveness. Thus, men are said to enter and remain in the armed forces for the same reasons they enter and remain in any civilian occupation. The focus of analysis for econometric models is on the concern for "hard" data, such as wages, amount of free time, paid vacations, working conditions, and status. The assumption is that the military does not represent or require a special calling, does not require a vocation, and is not meaningfully different from any other business. Moreover, it is assumed that what motivates men in the larger societal context—the pursuit of power, prestige, and income defined in terms of self-interest—also motivates men in the military environment. This assumed similarity of motivation permits the application to the military of a whole range of techniques that have proven useful in the organization of civilian business enterprises.

For the military, occupationalism is manifestly destructive of professionalism. It may well be true that certain concerns of the business executive are of equal concern to the soldier, although some available studies show they are not. But even if the same motivational forces initially attracted men to military service, it is unlikely that they could sustain men in the profession for very long. Econometric formulas cannot explain why men remain together under the terrifying stress of combat or why they are prepared to give their lives in the service of their comrades. Econometric models do not work on the battlefield; death in combat is always an uneconomic, irrational decision for the one who is doing the dying. It can be hoped that the movement toward occupationalism will ultimately be unsuccessful, but in the process it might succeed in transferring many of the inequalities of the larger economic order to the military itself. For example, the military normally treats men fairly equally; indeed, it has been in the *forefront* of removing many of the racial, religious, and ethnic barriers found in the larger society. The military was the first to racially integrate, and the military led the way for equal treatment of all soldiers. The utilization of econometric motivators to attract men to service may create a military that fills its ranks through conscription by hunger, poverty, race, and lack of opportunity. The result, in fact, has been the All-Volunteer Force which is blatantly discriminatory in terms of race and social status and whose socioeconomic composition is totally unrepresentative of the larger society.[20]

Occupationalism represents a threat because it will transfer the inequalities of the larger society to the military. Once military service is defined in essentially economic terms and motivation of the soldier is rooted in economic egoism, any notion of service that transcends individual self-interest is immediately destroyed. It is nonsense to suggest that the military can recruit and maintain men in service largely on the basis of economic self-interest and at the same time maintain that the military is a profession dedicated to selfless service of its clients. The fact is that the use of economic formulas for recruitment and retention fundamentally diminishes not only the ethical but also the professional supports of the military. A military profession cannot operate on the same motivational basis as a business, for the two work at cross-purposes.

Occupationalism is being forced on the military in an effort to save money and to analyze expenditures in terms of cost-effectiveness. At the same time, of course, the military itself has all too willingly embraced occupationalism in an effort to please its civilian overseers. In the end, the movement away from vocationalism represents a movement away from professionalism, and the movement toward occupationalism is a movement toward the destruction of the military profession as such.

Occupationalism can have potentially devastating effects on the ethics of the military profession. As the military moves away from status as a vocation toward status as an occupation, both the military's self image and the populace's image of the profession are beginning to change markedly. When the military views itself as a vocation, the populace can rightly expect adherence to high ethical standards. Moreover, the officers and men of the profession believe that their adherence to ethical standards, at least in principle, is what essentially separates them from the civilian society. There is a recognition that military service requires a sense of special obligation and responsibility. Thus, members of the military have their role legitimized in terms of their service to the profession and to the community.

With the acceleration in the movement toward occupationalism, however, the military has become more job oriented, and changes in the perception of the military role have begun to occur. The public still expects adherence to strict ethical standards, but military professionals are now treated as any other employees. The military is to be motivated by standards of self-interest, while at the same time the populace continues to see the military as performing selfless service. The difficulty arises when the military becomes perceived as just another occupation, when success in it becomes defined in terms synonomous with the values of the business executive, and when military professionals have to function in a marketplace environment more conducive to no ethics or situational ethics than to observing proper ethical codes. The fear arises that the profession will suffer an ethical breakdown[21] if it becomes indistinguishable from any other self-interested business enterprise. When the motivational, organizational, and promotional techniques of the military professional become identical to those of business, the decline of military ethics seems the natural consequence. One cannot destroy professionalism in the name of economics and at the same time expect that the military's sense of special obligation fulfilled in service will remain intact.

MANAGERIALISM

Occupationalism has been accompanied by managerialism, the penetration of the profession by a range of values closely associated with civilian business enterprise. It has been argued that if military service can be equated to a civilian business, many of the business corporation's organizational techniques and values can also be applied to the military. As secretary of defense in 1960, Robert McNamara believed that the application of business management techniques, such as cost-effectiveness, personal management, centralized promotion, computer purchasing, and salary control, could be applied to the military establishment without disturbing the essential quality of the military profession.

Over the last twenty years, many of the American military's traditionally successful mechanisms of control and organization have been replaced by mechanisms from the business corporation. There are several obvious examples of this replacement. Some of the traditional rights of promotion reserved to commanding officers have been removed and centralized at MILPERCEN. Leave and pass policies are no longer subject to disciplinary requirements by local commanders but rather are guaranteed by the central authority as a condition of employment. Soldiers no longer live in barracks but in two-man rooms, despite the fact that such accommodations have recently been noted to be among the strongest institutional supports for drug use among the soldiery.[22] Cost-effectiveness has become the requirement against which all other standards are measured; local commanders spend greater amounts of precious time filling out reports in order to satisfy bureaucratic requirements of the reporting system than with their men; and the establishment of a readiness system in which paper reports and surveys are substituted for actual readiness checks are all examples of the penetration of managerialism and managerial techniques within the military.

Among the most damaging aspects of managerialism has been the adoption of managerial *values*. The value of individual self-interest has made itself felt in the military profession, especially the officer corps. In *Crisis in Command*, I described a range of managerial values afflicting the officer corps that were directly traceable to the transformation of the military from a vocation into a business. These included careerism, situational ethics, "becoming a team player," lack of dissent, lack of resignations, persistence of unworkable policies in order to fulfill paper requirements, an officer corps in motion continually punching its tickets in order to qualify for promotion, and sending too many officers to staff school despite the inability to employ most of them in positions for which they are extensively trained.[23] At the combat unit level, one can even discern the tendency to equate captains and platoon leaders, the key killing combat commanders, with "middle-tier" managers. Central is the proposition that somehow military leadership can be replaced by managerial techniques.

Ever since the Vietnam War, the "management of resources" has been substituted for the leadership of men. The notion of resource management has become a substitute for leadership with disastrous consequences. The change in values within the profession has eroded the warrier ethos of leadership and has replaced it with managerial techniques, buttressed by the view that a manager in one area of expertise is equivalent to the military leader. Besides being contrary to the essence of the military profession, the difficulty is that the substitution of managerial values for professional ones fails miserably on the field of battle. Judging from the range of pathologies demonstrated by battle units during the Vietnam War, we can see that the adoption of the managerial ethos worked against establishing and maintaining cohesive combat units. Despite grotesque efforts to manage men to their deaths, many military managers found that few would go. The military began to internalize a number of managerial values, which continued to supplant traditional military values. The result in Vietnam was a battlefield catastrophe; the results for the future are likely to be equally ominous unless the impact of managerialism is stopped quickly.

CONFUSION WITH BUREAUCRACY

Confusing the execution of the responsibilities of the soldier with those of the bureacrat is inevitable given the tendency to view the military as no different from any other business. There is a fundamental difference in the values of the soldier and those of the bureaucratic manager, however. The bureaucrat is charged with executing rules that are deliberately designed to limit initiative and discretion. By executing rules "without respect to persons," the bureaucrat avoids responsibility and minimizes the necessity for judgment. The bureaucrat's ulitmate career goal is to follow orders and to minimize career risks, ever moving upward as he continues to manipulate the rules of the system to his own career advantage. The role of the soldier is radically different, even in today's managerial age. The core of military leadership is the ability and willingness to exercise judgment. If judgment is vital to leadership, then the soldier's code of ethics can never be the equivalent of mere bureaucratic rules. For the soldier a code of ethics represents a set of guiding norms encouraging judgment and initiative. While the bureaucrat seeks only to follow orders, the officer must apply judgment; where the bureaucrat shuns initiative, the officer must seize it; where the bureaucrat seeks to avoid responsibility, the officer must willingly accept it. What the military profession must understand is that, while it may be possible to make a good leader a good bureaucrat, it is impossible to make a good bureaucrat a good leader. The continued confusion of leadership with managerial bureaucracy remains a self-inflicted wound in the military profession.

With regard to the officer corps, bureaucratic organization and values will transform officers into bureaucratic functionaries unless a clear ethical code that supports military professionalism exists. Such a code must extol courage, flexibility, and the willingness to make decisions, take risks, accept responsibility, and exercise judgment. Such qualities have historically marked effective leadership and professional officership, and they are in as much demand today as they have been throughout history.

SPECIALIZATION

As the military's organizational structure became more complex, specialization became necessary. Colonel Richard Rosser, in the *Seaford House Papers*, maintains that specialization within the military has tended greatly to reduce professionalism.[24] He suggests that specialization works against commitment to the profession in the larger sense and substitutes for it a narrow commitment to one's area of special skill:

> Perhaps the biggest challenge to the concept of military professionalism is the need for specialization in all ranks. Young men in the service increasingly think of themselves as meteorologists, economists, electrical engineers, political scientists, and nuclear physicists. If they have a commitment, it is primarily to their particular profession or discipline and secondarily to the military profession.[25]

Military branches that allow the shorter career periods of twenty to twenty-five years exacerbate the effects of specialization. The men in these branches often must look beyond their period of military service for employment in order to sustain their families. As a consequence, they often seek to acquire skills within the military which they can use in the civilian economy after they leave active service. Specialization, both as a force within the military and as an anticipated reaction to leaving it, lessens the members' sense of commitment to the military profession as a way of life.

In itself, specialization has not necessarily undercut military professionalism. Given the amount of time a soldier spends in the profession, his career can be structured to offset the pull of whatever specialty he has. Although a soldier may have a technical skill, he still exercises that skill within a military environment, and one would think that that environment would be strong enough to remain his primary interest. In other professions, specialization does not seem to reduce commitment to professionalism. There is far greater specialization within the professions of law and medicine, for example, to say nothing of academia. Yet, members of these professions, despite, in some cases, very narrow specialties, have no particular problem identifying with the overall field.

A soldier does not necessarily spend a significant amount of his career assignments within his field of specialization. Indeed, a study of the typical successful officer who attains general officer rank would show that in a period of twenty-two years such an officer could expect to "punch through" approximately eighteen different assignments, spending very little time in any one of them. The military promotion system itself encourages the creation of generalists by rotating officers through a number of assignments rather than creating a core of specialists. One would think, then, that this system would help offset the destructive impact of specialization.

The fact is that specialization by itself is not an overwhelming force. It represents a major challenge only in combination with more virulent challenges that have weakened the profession. Specialization is having a great impact only because the profession has already been weakened by more powerful challenges.

Historically, the application of technology and specialization has seldom caused a weakening of military professionalism. The Roman Army, for example, had the most technologically sophisticated military machine of its time; yet, it was able to maintain a stunningly high degree of professionalism. The same is true of the Mongol armies of the fourteenth century as well as Napoleon's military forces. Thus, technology and specialization are harmful only when the military profession is unsure of itself and when it has already been weakened by other corrosive forces. Indeed, without specialization it is unlikely that a modern military force could sustain itself in warfare. The true challenge of specialization is not to rid the profession of it but to control it, and have it serve the goal of professionalism.

COMPETING ETHICS

Alternative sets of competing ethics are being adopted in the military. One of these is ethical egoism. Proponents of this notion argue that a community ethics transcending the interests of the individual within the community is not viable. In their view, all that is possible is the individual's ethics which is fashioned from the specific circumstances in which he finds himself at the moment. Among the most prevalent and accepted forms of this ethical challenge to military professionalism is the position known as existentialism.

An article published by a young junior grade lieutenant entitled "The Career Officer As Existential Hero" makes a case for a military existentialism which proposes an alternative ethics to form the basis of military professionalism.[26] The position should be explored in some detail because it provides a classic example of the view that philosophical egoism can provide the basis for community ethics within the armed forces. The author argues that such "old ideals" as duty, loyalty, and honor are dead for his generation of officers. He suggests that these concepts are "external commitments" that are forged in the community and superimposed upon the individual soldier. Since they are rarely if ever internalized by the soldier, they quickly collapse in the stresses of wartime. As an existentialist, he argues that allowing individual soldiers to develop a set of internal existentialist "commitments" will serve as greater motivation for them.

> I only ask that they not try to influence me into accepting their motivations. Truth, after all, is a relative quantity; the environmental factors influencing the graduates of the service academies of thirty years ago have changed; thus the motivational factors must also change."[27]

The logical consequence of this position is that since one cannot create a set of community ethics within the military profession that are different from those in the larger society, individuals within the profession need to adopt those ethical views congruent with the larger society. Thus, "what is required today is the projection of the image of the career officer consistent with the aspirations spawned by the contemporary society. Senior officers . . . must respect the claim to a different motivation by not trying to force their motivations upon us."[28] It is the individual alone who develops whatever motivations lead him to enter military service, and the larger professional community in which he claims membership may or may not share these motivations. In any event, the individual reserves to himself the absolute right to develop whatever values or goals motivate him. In addition, according to the existentialist, because man is a social being, those values that motivate him in any group are essentially those that will motivate him in the larger society. Thus, the author specifies that existentialist motivations for the soldier must be those which motivate individuals in the society at large.

Professionalism and the Brotherhood of Arms 103

The existentialist also suggests what the role of the military profession ought to be vis-à-vis the individuals who comprise it.

A true profession must provide the opportunity to work with and for people, the opportunity to influence others, the opportunity to master a discipline, and those aspects which make the military and the naval profession particularly appropriate for the fullfillment of an "existential commitment"—the opportunity to develop a level of personal and professional excellence and have it meaningfully challenged in positions, the opportunity to do battle with potentially overwhelming adversaries, and the opportunity to aid in the conquest of man's last two frontiers, space and the seas. Dedicated service to our country will still follow.[29]

Thus, the duty of the military profession is to provide *opportunities* for the individual and in providing such opportunities, the individual will confront circumstances in which he must make choices. It is in making these decisions that the existentialist finds meaning in the world and in his life. This is to say that the existentialist is what he decides; man is judged by the consequences of his acts, and a man is only what he decides to do. In this view, the role of professional standards is to provide opportunities within which the individual may make choices and come to define himself. The idea that the individual may have obligations to the profession that transcend the circumstances in which decisions may be made is, of course, unacceptable. Finally, it is assumed that once the individual decides what he wishes to do, dedicated service to our country will still follow. In essence, it is presumed that there is no fundamental tension between what the individual wishes to do in a given set of circumstances and what the military profession and professional ethics must compel him to do. It is in "doing," in deciding among circumstances, that the individual defines himself and thus renders service to the profession.

Several criticisms can be made of existentialism within the military setting. First, existentialism undercuts any notion of service to the community as a basis for a profession. While one may witness such service, it results only tangetially from the fact that the individaul "defining" himself in the larger professional or social context. There can be no question of having to sacrifice individual interests for community interests. Such a notion runs contrary to the very premises of philosophical egoism which rest at the base of existentialism.

A more specific criticism is that the argument for a military existentialism posits a distinct difference between *external* commitments and *internal* commitments. In this view, values imposed from without are not truly values at all; only those values that evolve from within are likely to be meaningful to the individual and compel observance. While no one would dispute the fact that internalized norms are excellent motivators, two points are worth consideration. The first is that in some instances external communal values may in fact be the only ones available or the only way to compel observance to necessary courses of ethical action. The second point is that there is an intimate connection (or ought to be) between external and internal values. Man is a social animal and he lives in socially organized groups. Much of what is meaningful to him is extracted from that membership; that is especially true with regard to the norms of the military profession which, as a rule, are not found outside its boundaries in the society at large. Individuals find meaning in their own personal lives largely as members of groups. Frequently, the external values of the group become internalized and the most effective stimuli for ethical action within the profession. Without a clear statement of external standards, especially ethical standards relevant to the profession's actions, it is difficult to see how an individual within the profession could readily arrive at these standards at all. Indeed, even if it were possible for an individual to arrive at these standards by intuition or, perhaps, even through experience, it is by no means certain that a community of individuals would end up discovering the same values. Even if one assumes that the values that develop internally are the best motivators, it is still clear that internal values will only develop within group settings and that they evolve more rapidly and more clearly insofar as the group sets certain "external" standards that must be observed as the price of belonging to the group. To deny this is to suggest that somehow, as if through intuition or magic, individuals within a community can spontaneously arrive at identical values. There is precious little evidence that this is so; it appears more in the realm of an act of faith.

Yet another critique of existentialism rests in its affirmation that "truth is, after all, a relative quantity." While this is not the forum for a debate as to what is truth and as to whether truth is an absolute or relative entity, the fact is that any profession must offer a statement of values and ethics that transcends its membership and defines what the profession is all about. In this sense, truth is not a relative quantity. If the profession does not profess an ethical center or set of obligations that constitute the "truth" of the profession, then, quite simply, it is not a profession. To suggest that such values are merely relative quantities within the context of professional ethics is to suggest that one professional may have grossly different professional values from another and remain a true professional. It implies that individuals within the profession may harbor different core values about what constitutes their obligations to the profession and what the obligations of the profession to the comunity are. For example, it is absurd to suggest that a medical doctor who feels comfortable with a set of values condoning the extermination of the aged should be allowed to implement those values simply because he feels they are "meaningful" to him. Nor would anyone expect him to be allowed to remain a professional in good standing. In the end, the question of whether or not truth is relative is not as important as the position of the philosophical egoist that values and standards have no meaning apart from the individual. The very foundation of military professionalism, as well as the very idea of ethics, requires that the values and ethics of a profession have a meaning in terms of the profession that is quite apart from the willingness of the individual to observe them. Only when the individual willingly observes such values is he truly regarded as a member of the profession. It is not the task of, nor is it possible for, the profession of arms to tailor its values to a multiplicity of individuals who have developed a wide range of conduct, even contradictory conduct, and still claim it is a profession.

With regard to the existentialist position that the values of the individaul are "spawned" by those of the larger society, men live in social groups, and the different social groups in which they find themselves can and do provide different values appropriate to action as a member of the respective groups. There is no absolute requirement that the values of military professionalism be inextricably tied to those of the larger society or that the motivational forces that work in the larger society work in any of its subgroups. Quite the contrary. A whole range of societal subgroups require motivational forces that do not work in the larger society but are still requirements for group membership. The affirmation that there is a fixed set of motivational factors that serve the individual *regardless of the social setting* or circumstances in which he finds himself is blatantly false. In its simplest form, some motivational factors are appropriate to some social situations and absolutely inappropriate to others.

Nowhere is the existentialist's philosphical egoism more evident than in the proposition that the function of a social group, in this case the military profession, is to provide a series of opportunities to which individuals may react. In reacting to these opportunities, individuals make decisions that define "who the individual is." The stress on opportunities obscures the fact that the military profession has far more to do with *obligations* than it has to do with providing opportunities. To suggest that the *raison d'etre* of the military profession is to create opportunities for individuals to react to is to negate any ethical standards as to *how* men ought to act since the existentialist holds that the challenge comes from responding to circumstances in which men find themselves at the moment. It is important that the military profession evolve standards of how men ought to act in order to guide them when they have to act. The existentialist holds that men know how to act either through intuition or through the relevant information obtained in the very circumstances in which they find themselves ready to act. If the only basis for ethical action is the data provided by the circumstances themselves and if every set of circumstances is unique, one learns nothing from one's own experiences. For the existential hero, there is no basis at all for obligations since obligations would bind the actor *before* the circumstances in which he must act occur. Thus, the fundamental thrust of existentialism is to deny the ability of men to generate ethical rules that guide human conduct. As a corollary, it is impossible to devise any ethics or moral standards to guide and measure human action within the confines of a profession. Existentialism as an alternative ethical perspective is destructive of military professonalism, as it is of any sense of obligation apart from those which the individual discerns to be imposed by the circumstances of the moment.

All five challenges to military professionalism have one thing in common; the underlying proposition that the individual pursuit of self-interest somehow leads to the highest form of human fulfillment. With regard to individuals in group life, the assumption is that the pursuit of self-interest will necessarily produce a sense of community interest shared by all. The corollary is that the pursuit of individual self-interest can never truly be contrary to the good of the community. This same set of propositions forms the foundation of most Western capitalist democracies. The difficulty with them is that they are open to serious question from a variety of perspectives. The fundamental problem is that one cannot possess a sense of community obligation for the military profession that will not at some point require the individual soldier to forego his self-interest in order to secure community interests. At the same time, one can never form a community of military professionals in which community obligations bind the soldier apart from his self-interest if one assumes that individual self-interest is the highest operant value within the profession. All of the challenges to military professionalism come together with great impact to destroy the very center of the profession itself.

A profession that relies upon occupationalism, managerialism, and the values of business, that confuses itself with bureaucracy, that allows pyschological egoism to become its philosophical base, and that permits specialization to cut across communal ties may well have ceased to be a profession. What it has become is an entrepreneurial enterprise whose primary use is as a vehicle for fulfilling the career aspirations of the individuals within the profession. More to the point, whenever the entrepreneurial organization levies obligations or penalties upon individuals who are not observing their obligations, the response is to leave the organization on the pragmatic grounds that it is no longer serving one's interests. Basically, the tension is between an organization with a developed sense of corporate values and one with only mechanistic values.[30] The two are opposites and irreconcilable, especially in the profession of arms where responsibilities are so serious and the requirements for ethical action so obvious.

Responding to the Challenges

Responses to these challenges fall into two categories. First, there is the convergence response which suggests that the military can remain an effective military force only if it brings its values into line with those of the larger society. Second, there are those who would like to return to a "golden age" in which the military profession was either totally isolated from the society, or, in a curious corollary, one in which the values of the society were so supportive of the military that a strong congruence existed between the values necessary to maintain an effective military force and those operant in society as a whole. Neither of these responses seems appropriate, or even workable, considering how the challenges operate in modern society.

To those who would return to a so-called golden age of public support, it must be pointed out that in no period of U.S. history have the values of the larger society been congruent with and supportive of those values required to build an effective military force. From the very beginning of the republic, after all, there was a fear of the military. Civilian antagonism to the military was reflected in the views of the Founding Fathers themselves who feared that any sizable ground army might be easy prey for an executive run wild. They did support a sizable navy which would be at sea and therefore poorly positioned to become a force of consequence in domestic struggles.

In the United States, the military has always had to confront the antagonism of the larger society. According to Russell Weigley:

> Historically, in the United States democracy, the army and its values have tended consistently to seem so alien to the rest of society that for the army the times have almost always been troubled.... The tensions between army and society have been great enough that, for American soldiers attempting faithful service to the values of both, even dilemmas of moral integrity are not altogether new or merely related to Vietnam.[31]

One aspect of the tension between military and civilian society is found in the writings of Colonel Emery Upton in which he charged openly that the military profession was being perverted as a profession largely because of bad civilian policies; this was in 1912.[32] Within the United States there has almost always been an hostility of sorts between the larger civilian society and the military. Again to quote Weigley, "When military men lament that they find themselves in an inhospitable society, they will draw from American history only the small consolation that it has always been thus."[33]

The military in America has always been outside the societal mainstream, and the populace has generally failed to give the soldier the same respect for his profession that members of other professions have commanded.[34] For most of U.S. history, military officers and soldiers have had to set their own standards and values in the face of a society whose values were patently contradictory of military values and often openly opposed to the very idea of a standing military force. Thus, to want to return to an age when the military enjoyed the respect of the civilian populace is to pursue a myth. In fact, there was probably never such an age at any time in any Western society. To be sure, usually during wartime, heightened support for the military can be discerned.

The lack of a stable tradition of support for the military profession is not surprising considering that Western capitalist democracies are predicated on the pursuit of economic self-interest. Liberal capitalist democratic polities have been unable to generate values congruent with the profession's notions of sacrifice and service perhaps because to do so would require the militarization of the state more than the liberalization of the military. Moreover, in earlier eras military professionalism was not the result of a congruence of values with the larger society; exactly the opposite seems to have been the case. The military sustained and developed good habits of officership, cohesion, and discipline within the profession *despite* the fact that the larger society did not support its values.

Another strategy for responding to the challenges of military professionalism is to suggest that the military can engender within itself a sense of ethical bearing and military effectiveness by ensuring that its values converge with those of the larger society:

Military professionalism must ultimately be grounded on the premise that military ethics converge with the ethical values of the larger society. A military system in a democratic society cannot long maintain its credibility and legitimacy if its ethical standards significantly differ from the civilian values of the larger society.[35]

It is argued that when the values of the military profession are brought into line with those of the larger civil society, the result will be a profession secure in its own sense of service and obligations and more effective in carrying out its responsibilities. What is the evidence to support this view?

If one is going to talk about the convergence of miliary and civilian values, one must clearly specify the terms of the problem. If one is addressing the values of the society as they are meant in the generic sense and their relation to the military profession, it is obvious that some degree of congruence always exists. To the extent that the military's very existence only makes sense in terms of its special service to the state, Constitution, and society, there already is a convergence of fundamental values. No democratic society could (or ought to) tolerate for long a military establishment that did not see its primary role as service to the civil order. In this sense, to suggest that convergence is required to sustain a sense of military professionalism may be stating the obvious, that is, to require that the profession be a profession.

On the other hand, if by sustaining military professionalism through convergence one means that the values operant on a day-to-day basis of the larger civil society should be transposed into the military, one is making a serious error. In the history of the West, most civilian societies have been fundamentally antithetical to military professionalism. Such social orders are antithetical as regards not only the large goal of self-interest as the highest social value but also many of the routine actions which simply work against the maintenance of habits, values, and traits crucial to military effectiveness. If the strategy of convergence implies that the military must bring its day-to-day habits, traits, and values in line with those permitted in the civilian society, the price of convergence is likely to be a military profession that is without a sense of true professionalism.

Convergence as a strategy for responding to the challenges of military professionalism is really a two-edged sword. If one democratizes, liberalizes, and civilianizes the military, the real fear arises that the military will be unable to carry out its responsibilities in battle. Indeed, from what we know of cohesion and military effectiveness, it seems clear that many of the habits, values, and norms of the civilian society will not work in the military environment. Accordingly, convergence runs the risk of producing a military force whose values are, in fact, congruent with larger social values but unable to effectively carry out its mission. Convergence could easily erode military professionalism by eroding the military capability of the profession itself.

An even greater difficulty is implicit in the strategy of convergence. If the profession were to conclude that it was being weakened by the importation of civilian doctrines that were reducing its combat effectiveness, the profession might face a truly hostile dilemma: the choice of silently watching the profession being destroyed in a sea of self-interest, or deciding that the larger society was a threat to the very existence of the profession itself. From this perspective, the danger of the doctrine of convergence is that it lays out the logical rationale for a strike at the state by the military in an effort to reform the state. If the only nexus of military professionalism rests in society's larger values, that is, if no military force can be effective unless it adopts civilian social values, then when the military finds itself to be ineffective the temptation to reform the society as a way of reforming the profession may become irresistible. The strategy of convergence lays the groundwork for just such a temptation.

The assumption that military professionalism and effectiveness will somehow follow from a convergence of military and civilian values is an act of faith for which there is little evidence. Actually, the evidence suggests that exactly the opposite will occur. Military structures are quite capable of remaining intact as professional institutions and of conducting effective combat operations, in the face not only of a lack of support from their larger societies, but also sometimes outright hostility to them. For example, the Roman legions cohered long after Roman society had ceased being the moral republic that supposedly gave the legion its strength. The German armies of World War II continued to fight long after German society had been pounded to dust; the German armies of World War I displayed the same cohesiveness. The ability of French troops to endure at Dien Bien Phu strongly suggests that the support of the larger social order is not necessary to the operation of an effective military force. None of this suggests that such support is not desirable; it is only to suggest that the military profession is so different from the civil society, requiring such a different set of special obligatons and responsibilities, that it is unrealistic to expect that societies rooted in self-interest can meet the requirements of military professionalism by forcing the profession to bring its values into line with those of a self-seeking society. The reality is that the military must, at times, be the guardian of its own values, codes, and discipline that produce military professionalism and, ultimately, combat effectiveness. To be sure, the military must never forget that its own honor and professionalism are rooted in service to the state and its civilian masters. There is no evidence, at least in most Western societies, that a necessary contradiction exists between a sense of military professionalism and the larger democratic values of the civil polity.

Avenues of Advance

At least three directions of professional development seem possible for the military. The first posits that the military profession has historically been isolated from the larger social values of its host society. Again, this is not to suggest that the profession does not share the generic values of the society; the *raison d'etre* of the profession is its service to the state. But the values associated with an effective military simply cannot be the same values of a civil society founded on self-interest. Some degree of separation of the military from the larger society's values is a basic condition for rebuilding and sustaining military professionalism.

A series of studies conducted on the modern military reveal that soldiers, especially officers, tend to reflect the general socioeconomic characteristics of the larger society.[36] At the same time, they have the distinct perspective of their profession. Moreover, they see themselves as having to make higher sacrifices than the average citizen, as well as being the repository of the best values of the republic. The point is that being *in* a society does not necessarily mean that the solider must be *of* it. Rather, being a member of the military profession seems to enforce a degree of isolation from the civil society, and this isolation allows professional values to take hold. Some soldiers may well feel that the civilian populace has low esteem for their status, but, to the contrary, a series of national opinion polls have shown that the military remains one of the most prestigious institutions of all in the eyes of the general populace.[37] Indeed, the *civilian* population tends to see the military as both separate and special.

On balance, public support for the military today is about the same as it has always been. Even during the Vietnam War support for the military never fell very low in the eyes of most citizens. If the military is to respond to its challenges, one of the basic requirements is that it remain a profession, and so it must remain somewhat apart and distinct from the larger society in its values.

Two analysts, Clotfelter and Peters, in a study of the effect of civilian-military interaction on the attitudes of military professionals, found that increased interaction tended to have no effect on reducing the sense of separateness and professionalism felt by the soldiers. When hostile interaction occurred, the effect was to reinforce the professional's sense of being special.[38] They also found that officers see themselves as a distinct social group, as upholding the best traditional values of society, as requiring and giving greater dedication than civilians, and they see their profession as involving greater sacrifice than other professions.[39] What the data imply is that to some extent, the military profession is already separated from the larger social order and that the separation serves the profession well. That separation seems to be specific and to apply only in those areas where the military professional feels he must be removed from the civil society in order to be true to his profession.

A second direction for military professionalism has to do with the source and condition of the profession's core values. If many civilian values are not conducive to the development of military professionalism, what is the source of the profession's values? Historically, the source has been the profession itself.

One reason why the challenges to military professionalism have been so effective is precisely because they have supplanted a number of traditional values. The military must realize that the profession itself must preserve its uniqueness, just as the medical profession must preserve its own professionalism by setting standards and enforcing them. The military cannot expect the larger society to provide the values appropriate to its profession.

Most of all, the military must be prepared to defend its core values from assault from without and erosion from within. One of the major criticisms of the military is its failure to control the impact of external social forces on its members.[40] History is replete with examples of military structures hammered by antithetical values and habits spawned by their civil societies. Successful professions have formed limits around them that institutionalize, mitigate, and control negative forces. The American military over the last two decades has failed to protect itself, acclimating itself too readily to civilianization, occupationalism, and managerialism. The military must come to understand that it and not the civil society bears the primary and direct responsibility for developing and sustaining the values necessary to its own professionalism. All the while, it must recognize that its role requires service to the larger society. The task is not easy, but certainly not impossible.

For those who are concerned lest the military become too isolated, it should be emphasized that a thoroughly professional military is far more easily controlled than one that closely reflects the momentary values of the larger society.[41] If the military were to mirror the values of the larger society, when the civilian populace clamored for a certain course of action or took to the streets, these same forces would create an undisciplined mob of uncertain loyalty within the military. Reginald Brown states that the development of a strong sense of military professionalism clearly eliminates a range of action that the military can legitimately undertake.[42] One of these is, of course, responding directly to the winds of popular political change. Thus, military professionalism is not a threat to civilian control as much as it is a force for confining the overt *political* response of the military to very narrow channels.

The danger of praetorianism in the American context is greatly exaggerated. In a number of military establishments in the West, professional and aristocratic military professions easily exist within the context of democratic societies without presenting any real threat to civilian control. England represents a classic case as do the United States, Canada, and Israel, and generally the French as well.[43] Apparently, there is no inherent contradiction between a liberal democratic civil order and the presence within it of a professional and authoritarian military establishment. Both attempt to do different things and require different sets of values.

Finally, the third direction of military professionalism is to develop a code of ethics that clearly defines what the profession of arms is all about. Without such a definition, the military risks becoming like any other civilian occupation and not having any ethics at all. The codification of a set of ethics also requires that those who cannot or will not meet the requirements must leave the profession. The military is not for everyone. Professionalism means living up the the requirements and expectations and carrying out obligations that define membership in a group laying claim to a special task.

The term "military professionalism" is often misundertood. It is commonly confused with following rules, orders, and "doing one's duty." *Duty is the servant of ethics; it is not a substitute for it.* An example of the confusion is seen in the following statement:

> The professional is susceptible to influences external to the profession, albeit in a less pervasive way. Buffeted by these various forces, the individual professional is nevertheless expected to follow the lifestyle and accept the morality and ethics of a profession that primarily evolves from a monastic focus and horizontal network. Although the profession operates within the context of the morality and values of the political-social system, these dilemmas are relieved by adjusting individual lifestyles to the expectations of the profession. Thus the perspectives of the profession become the dominating morality and ethics of the individual officer. Institutional articulation of integrity, duty, honor, country and officership are substituted for the individual's own sense of morality and ethics.[44]

This position misinterprets the role of ethics in a profession. If professional ethics is substituted for an individual's total sense of ethics, the military professional risks becoming a value-free technician willing to apply his professional skills regardless of ethical consequences. Professional ethics never constitutes the sum total of an ethical man. It is most relevant to the exercise of his role within the military profession. To be certain, professional ethics can never be *malum in se*, but the individual soldier draws values from a variety of group memberships. At best, the ethical values of the profession are adequate guidelines in situations relevant to the military profession itself. Even then the individual cannot legitimately abandon his ethical self. If he does he becomes ethically deformed. As noted earlier, the observance of obligations without understanding, choice, or a sense of their consequences is not ethics or duty; it is a perversion of both.

A code of military ethics would not constitute the total sum of an individual's moral being or his obligations. As General Maxwell Taylor has said, "setting forth the principles and standards of professional behavior to guide the deportment of military leaders...would not presume to serve as a universal ethic for all men at all times or even for officers for fulfilling obligations unrelated to their profession."[45] Development of a code of ethics will help a great deal but will not, in itself, solve the ethical problems of the military profession. At least two additional elements are required. First, the soldier must be given adequate training to develop the ability to recognize ethical dilemmas that are likely to arise in the conduct of his profession, and, second, he must develop a judgmental ability to resolve those dilemmas in an ethical way. If this means, as it often does, choosing between the obligations of his profession and those acquired elsewhere, then that is the very nature of ethical choice. One cannot escape ethics by substituting obedience for them; to do so is to pervert ethics.

The military and all other professions must realize that tensions may arise between an individual's sense of ethics and his relationship to his profession, or any number of possible sources of ethical obligations. Man lives in a less than perfect world. The problem is not to bring the values of the individual into perfect symmetry with those of his profession; the problem of ethics is for man to be able to choose correctly among conflicting obligations when circumstances will not permit him to observe all his relevant obligations. He cannot truly escape this choice and yet remain an ethical agent. The key to a true military professionalism, to building a wall to resist the penetration of hostile values, is to create, promulgate, and enforce a code of ethics, and then train the military professional in the ways of ethical reasoning. Such is the challenge which the military must meet in the next twenty-five years.

Conclusions

Life and service are qualitatively different in the military from other occupations in civilian society. The profession of arms is in itself categorically different from other professions. If the military gains its legitimacy from the recognition that its reason for being is to serve and protect the civil order that is its client, there can never be any question of a military professionalism that transcends the society that bestows legitimacy upon it as long as the civil order itself does not, by its actions, become *malum in se*. The military profession has been subjected to a series of external challenges to its professionalism. These challenges have been inordinately effective in reducing the professional character of the military partly because the military has failed to control and channel them into acceptable areas. The result is a military profession whose professionalism is in serious question.

Future challenges require that the military devise developmental strategies that can control and reverse some of the pathologies that have grown within it. This analysis suggests at least three directions of professional development. The military must understand (1) that in order to sustain its professional stature, it must keep a certain distance from the civilian way; (2) that it must truly be the master of its own house with regard to generating values appropriate to its professionalism; and (3) that the very center of the profession requires the evolution of an ethical code that clearly defines the obligations of its members. Sooner or later, the profession that starts on the road of convergence will become a mirror image of the larger society—in which case it will be a profession no longer.

Notes

1. R. I. Aitken, "The Canadian Officer Corps, The Ethical Aspects of Professionalism," *Canadian Forces Staff School* (unpublished paper, April 1979), p. 20.

2. John Keegan, *The Fact of Battle* (New York: Vintage Press, 1976).

3. Alan Lloyd, *War in the Trenches* (New York: David McKay Co., 1977).
4. Edward A. Shills and Morris Janowitz, "Cohesion and Disintegration in the German Wehrmacht in World War II," *Public Opinion Quarterly* 12 (1948).
5. Samuel Stouffer, *The American Soldier* (Princeton, N.J.: Princeton University Press, 1976); see also Richard A. Gabriel and Paul L. Savage, *Crisis in Command: Mismanagement in the Army* (New York: Hill and Wang, 1978).
6. Arthur J. Dyck, "Ethical Bases of the Military Profession," *Parameters* (March 1980):39-46.
7. Ibid., p. 44.
8. Lewis Sorley, "Competence As an Ethical Imperative: Issues of Professionalism," Paper presented at the IUS Regional Conference, Maxwell Air Force Base, Alabama, June 4, 1979, p. 3.
9. Dyck, "Ethical Bases."
10. Samuel P. Huntington, *The Soldier and the State* (Cambridge, Mass.: Harvard University Press, 1967), p. 67.
11. Sorley, "Competence," p. 4.
12. Sir John Hackett, *The Profession of Arms* (London: Times Publishing Co., 1962), p. 63.
13. Ibid.
14. Robert G. Gard, "The Military and American Society," *Foreign Affairs* (July 1971):699.
15. Walter Kerwin, "The Values of Today's Army," *Soldier* (September 1978), p. 4.
16. James H. Toner, "Sisyphus As a Soldier: Ethics, Exigencies, and the American Military," *Parameters* 7, No. 4 (1977):4.
17. Ibid., p. 5.
18. Ibid.
19. Kerwin, "Values of Today's Army," p. 4.
20. For a thorough analysis of the socioeconomic composition of the AVF, see Richard A. Gabriel, "About Face on the Draft," *America* (February 9, 1980): 95-97.
21. This term is taken from an article by Allan J. Futernick, "Avoiding an Ethical Armageddon," *Military Review* (October 1979):17-23.
22. Larry H. Ingraham, "Drugs, Morale, and the Facts Of Barracks Living in the American Army," paper delivered at the Thirteenth Annual Anglo-American Psychiatry Symposium, Plymouth, England (October 1978); for more on this point, see the larger official study by the same author, "The Boys in the Barracks" (Washington, D.C.: Walter Reed Army Institute of Research, 1978).
23. Gabriel and Savage, *Crisis in Command*.
24. Richard R. Rosser, "Civil-Military Relations in the 1980's," *Seaford House Papers*, 1970, p. 68.
25. Ibid.

26. David G. Deininger, "The Career Officer As Existential Hero," *U.S. Naval Institute Proceedings* (November 1970):18-22.
27. Ibid., p. 22.
28. Ibid.
29. Ibid.
30. For further discussion on this distinction, see Richard A. Gabriel, "Acquiring New Values in a Military Bureaucracy: A Preliminary Model," *Journal of Political and Military Sociology* (Spring 1979):89-101.
31. Russell F. Weigley, "A Historian Looks at the Army," *Military Review* (February 1972):26.
32. Ibid., p. 33.
33. Ibid., p. 34.
34. Ibid.
35. Sam Sarkesian and Thomas M. Gannon, "Professionalism," *American Behavioral Scientist* (May-June 1976):506.
36. Franklin D. Margiotta, "A Military Elite in Transition." *Armed Forces and Society* (February 1976):155-185.
37. Ibid., p. 165; "What America Thinks about Itself," *Newsweek*, December 10, 1973, p. 45; Jerald G. Bachmann and Jerome Johnston, "The Volunteer Armed Force," *Psychology Today* 6 (October 1972):113.
38. James Clotfelter and B. Guy Peters, "Profession and Society: Young Military Officers Look Outward," *Journal of Political and Military Sociology* (Spring 1976):42.
39. Ibid., p. 45.
40. Gabriel and Savage, *Crisis in Command*.
41. Reginald J. Brown, "The Meaning of Professionalism," *American Behavioral Scientist* (May-June 1976):520-522.
42. Ibid.
43. There have been a number of mutinies within these armies over the last two centuries. For the most part, however, these mutinies have not been designed to replace legitimate civilian governments as much as to oppose specific policies, so that even in the incidence of mutiny these armies have not represented a threat to the existence of the democratic state. The major exception is, of course, the French Army's attempted coup in 1958, but even here the circumstances were such as to create a state of civil war prior to the attempt.
44. Sarkesian and Gannon, "Professionalism," p. 510.
45. Maxwell D. Taylor, "A Professional Ethic," *Army* (May 1978):19.

5 A CODE OF MILITARY ETHICS

A number of arguments have been raised against the formalization of a code of ethics, and each is addressed in some detail in this chapter. Many of these objections appear to misunderstand the nature of ethics and ethical judgment, and the relationship between obligations acquired within and outside a profession. Some of these difficulties have already been addressed in the preceding chapters, especially in Chapter 2. Nonetheless, it is worth examining the question of whether a code of ethics for members of the military profession ought to be developed. It is the clear position of this work that an ethical code *must* be formulated if the profession of arms is to be able to meet the challenges of the next two decades and to resist any further erosion of its sense of special calling.

Advantages of a Code

With regard to the formalization of a code of ethics for the military, General Maxwell Taylor wrote:

There may be justification, or even a definite need, to restate in strong and clear terms those principles of conduct which retain an unchallengable relevance to the necessity of the military profession and to which the officer corps will be expected to conform regardless of behavioral practices elsewhere.[1]

General Taylor recommends a rudimentary code and emphasizes that the code would guide the *professional* behavior of members of the military and "would not presume to serve as a universal ethic for all men at all times or even for officers in fulfilling obligations unrelated to the military profession."[2]

Major General William Lynn also recognized the need for a code in 1971 when he called for the creation of a professional ethics board to establish a set of professional ethics and create a system for enforcing these standards.[3] In addition, as we have seen, every major study on the subject of ethics undertaken by the profession itself sees a need for a code of professional ethics. Those within the profession show a clear desire to establish a formalized code of military ethics.

What would be some of the advantages of a formalized code of military ethics? Lewis Sorley has noted that ethical conduct lies at the core of the trust between civilian society and the profession of arms. If the civil society is not convinced that the military guides its conduct by a clear set of ethical standards against which it will be held responsible, the whole notion of trust between civilian life and the military profession breaks down.[4] Sorley quotes from the Cadet Honor Code at West Point which notes that "during the long history of the profession of arms, strict adherence to professional, ethical, and moral codes has been essential if the power and influence of the military organization were to be an effective servant, rather than the arbitrary master of the state."[5] Thus, the formulation of a clear code of military ethics represents the foundation for a special trust between the civilian society and the military establishment that has sworn to protect it.

General Taylor cites several other reasons for constructing a formalized code of military ethics. He notes that a code would unify the profession in its shared respect for common ideals.[6] It would help create a sense of community by specifying the "cost of belonging" to that community in terms of the obligations that members of the profession are expected to observe. Without a common ethical center, there can be no profession. The creation of a community within the profession requires a special calling and uniqueness and should be formalized in a code of special ethics for all to see. At the same time, General Taylor suggests that a code of military ethics would help develop "character formative" forces within the profession.[7] It would set standards of socialization for the soldier and ideals towards which he could strive. Students of military ethics realize that perhaps not all ethical ideals are attainable without great effort. It is the striving itself that ennobles one. A code would, therefore, at least contribute to setting standards of character toward which the military could hope to socialize its new members.

A formalized code would also show the civilian populace that the military does indeed possess a special sense of obligation. It would demonstrate in unequivocal terms the pledge of the military to uphold certain core ideals. The creation of a specific code of military ethics would likely help promulgate ideals that would help solidify public support for the military. General Taylor suggests that a code would help in another way. Since the military must constantly interact at the higher levels of command with its civilian overseers, who themselves are often ill-equipped to deal with the military and few of whom are truly aware of the special nature of the military task, the formalization of a code of military ethics stating the special obligations of the military may help civilian superiors realize the nature of the military way and the background ambience necessary to it. Perhaps this will facilitate understanding between military men and their civilian superiors, and help civilians carry out their responsibilities with regard to directing the military.[8]

Finally, if the military has special obligations and expects to attract and keep good men to carry them out, it must understand that the way of the soldier is not for everyone. By clearly stating the special obligations and responsibilities of the profession of arms, by clearly stating the special sense of dedication and sacrifice required for those to serve honorably within the profession, the code will attract those men who are willing to live up to its stated ideals no matter how difficult. There is a false security in attracting adequate numbers of men to the military who believe that the military is merely another job. By stating what the military profession is and what it expects of its members, a code can be a powerful force in attracting to the profession the very best in society, those who are willing to accept the special challenge of military service with all its hardships, sacrifice, and awesome responsibilities.

For these reasons and others, the formalization of a code of military ethics will help the military profession regain its sense of uniqueness. To quote General Taylor again:

> After surveying the many facets of this issue, I conclude that it is worth the effort to undertake the formulation of an officer's code, possibly as a first step toward one of wider scope for the entire military establishment.... It would proclaim to the world what the military profession stands for and by what standards it accepts judgement.[9]

A code of military ethics would be analogous to a book of monastic rules. It would specify the nature of the profession of arms, the obligations required of its membership, the price of belonging, and, indeed, the penalties for failing to observe the creed. The ethical imperatives should be phrased in unambiguous language. Some obfuscation and lack of clarity may serve the *law* well, but not the cause of *ethics*.

Objections to a Code

In some quarters there is substantial opposition to the formalization of military ethics in the form of a code of behavior. In order to support the establishment of such a code, it is important to examine some of the arguments advanced against a code. It is hoped that an analysis of these arguments will demonstrate the need for a formalized code by removing the most common objections to it.[10]

1. An ethical code for the military would be meaningless since one cannot teach ethics to begin with; rather, one must acquire ethics as a consequence of one's total life experiences. By the time one enters the military, so the argument goes, one's sense of ethics is already solidly set. In response, it can be noted that the proposition that that one cannot teach ethics is nonsense. If ethics is defined as statements of what one ought to do as a member of the military profession, clearly no one is born with any professional obligations, any more than one is born with a knowledge of physics. All ethics must be taught like the knowledge attendant to any discipline; all virtues must be learned as well. On the face of it, then, the question is not can one teach ethics, for the answer is yes because there is no other way to acquire ethics. The real question is how best to teach ethics.

Even though soldiers acquire much of their general ethical sense long before they enter the military profession, they learn almost nothing about the rules of ethical action *specific to the military profession*. Some general ethical precepts may be similar, but the specific ethical requirements of any profession must be learned after one has gained membership in it. A profession's use of ethical codes in order to teach the ethical obligations attendant to the profession is a necessary requirement. Indeed, even a good man who makes ethical judgments outside his profession and demonstrates certain virtues may have to learn new ethical obligations and acquire new virtues when he becomes a member of a special profession. No tool is more valuable to this task than the specification of a formal code of ethics.

The idea that a profession cannot change the ethical habits of individuals learned prior to entry misses the point. The fact is that no one has *any* ethical sense of a profession until he joins it and is made specifically aware of its ethical requirements. Moreover, even a good man outside the profession may not be a good man within it. A whole range of ethical values acquired outside the profession may be good for some circumstances that arise outside the profession but may at the same time be very poor guides for action within the profession. At the extreme, a Christian pacifist may be a moral man in many ways, but he is likely to make a poor military professional, especially if he is in command of a combat platoon. One ought not to confuse the possession of certain virtues with ethics nor assume that all virtues or ethics are particularly relevant to a given profession. To put the matter another way, some obligations and responsibilities are more relevant to some professions than to others. If by ethics is meant a series of obligations that specify what one ought to do in a profession, including the capacity for moral reasoning, then both must be taught after an individual becomes a member of his profession. This is especially so if the capacity for moral reasoning is to be relevant to the ethical judgments that must be made within an environment most attendant to the profession itself. From this perspective, to argue that one ought not to develop a code of ethics because one cannot teach ethics is to fail to understand the nature of ethics and how one acquires ethics. One cannot teach ethics without specifying the obligations necessary to ethical behavior, and there is no clearer way to specify those obligations than to enshrine them in a formal code.

2. Ethics cannot be enforced from without. Codes will remain meaningless unless ethical standards are internalized by the individual soldier. There is no doubt at all that ethical codes bind more strongly when they are internalized. But whether or not a formal code of ethics is internalized, a code can still provide a standard of judgment for professional action. Without a code of ethics, one can at least argue that it will be much more difficult to internalize professional ethical precepts because the precepts and their meaning are less clear. If the internalization of ethical norms is the goal, it must be clear as to what ethical precepts are to be internalized. And the best way to do this is to use a formal code. No one is suggesting that a code per se compels ethical behavior. What a code can

do is to specify those obligations to be learned, practiced, and reinforced. Recent studies have shown that the mere existence of a formal code of ethics within an organizational setting does indeed raise the level of ethical action if for no other reason than it clarifies what is expected as ethical conduct.[11] If one seeks the internalization of ethical precepts, in order to internalize values and norms one must first specify them by stating the precepts of a code. The mere statement itself is likely to contribute to the internalization process. To suggest that internalized values somehow negate the influence of a formalized code is to misunderstand the process through which values are internalized. One cannot internalize obligations that one is unaware of and one can be made aware of professional obligations very quickly by formally stating what they are in a code of ethics.

3. A code might become a substitute for ethical judgment. It is argued that minimum standards of ethical behavior have a way of becoming maximums. A formal code of military ethics might be perceived as constituting the total sense of ethical obligations for members of the military profession. Individuals would then live up to only a minimum code of ethical obligations which they would regard as the sum of their ethical responsibilities.

Members of the profession must be made to understand that a code of military ethics constitutes only a minimum set of obligations. A formal code does not constitute the sum of an individual's ethical responsibilities. Indeed, the obligations of the code may at times conflict with a soldier's other obligations. When this happens ethical choice is required. Moreover, the tendency to reduce ethical codes to legal rules distorts their function. As Malham Wakin has pointed out, ''the immature or unsophisticated frequently narrow their ethical sights to the behavior specifically delineated in the code so that what may have originally been intended as a minimum listing becomes treated as an exhaustive guide for ethical action.''[12] Any tendency to regard a code of military ethics as an exhaustive list of the soldier's ethical obligations is incorrect. But to say that men can pervert codes is not an argument against their existence per se. Ethical codes are general statements, and they usually delineate obligations in general terms. They seem to resist great specification. The function of ethical judgment is to determine what obligations are to be observed under what circumstances. To argue that codes may become maximum lists of obligations is to pervert them, but it is not an argument against their creation in the first place. The existence of a code per se will do little unless soldiers are also educated in moral reasoning.

4. A code would state ethical obligations in an ideal form, and many of the ideals would be empirically unattainable. Accordingly, there is no point in stating ideals that cannot be attained; such codes are useless. This argument misses the point of ethics and what it is supposed to do. As noted earlier, a basic philosophical premise for understanding ethics is the dictum "ought implies can." One could not legitimately establish an ethical code that set up unattainable ideals as their basis for action, for who could be held responsible for ethical obligations that could not realistically be observed in the empirical world?

On the other hand, the ethical obligations contained in a code of ethics should be stated in such a manner that they are difficult to live up to. A minimal code of ethics which everyone can observe all the time without any real effort is not a moral code at all. For a code to be useful in engendering ethical action, there has to be some gap between the ideal and the real, between aspiration and attainment.[13] This gap produces a kind of "creative tension," so that striving for ideals also makes an officer "good."[14] This concept derives from the Aristotelian notion of virtue, which states that striving for ideals tends to ennoble the individual regardless of whether he attains them. From this perspective, it can be observed that failure to live up to a code is not a criticism of the code at all as long as the precepts contained within it are attainable. On the other hand, if the gap between aspiration and attainment is too wide, at some point men will stop trying or begin to pay lip-service to codes they cannot realistically observe.

5. It is impossible to construct a code of ethics in the first place because the range of alternatives it would have to address would be impossibly large. Codification would also require the impossible task of all officers to learn how all precepts would apply in all circumstances, and enforcement of such an extensive code would be impossible. The argument suggests that a code of military ethics would have to be so specific as to codify in advance all the instances and circumstances under which any stated ethical precept might apply. The development of a workable code of ethics for the military profession would not then be possible.

This legalist argument tends to confuse a code of ethics with a body of law. The fact is that ethical codes are not the same thing as laws. Ethical codes specify what men ought to do, and the individual chooses which obligations he will observe when he cannot observe all relevant ones. The central point of a code of ethics is the necessity for choice. Law, on the other hand, obviates choice. The very specificity of law removes choice and substitutes obedience for obligation. Moreover, there is no necessary connection between a body of law and any specific ethical content. Most laws do not address ethical conditions at all. Speeding laws, for example, do not involve ethical conditions, nor do zoning codes. Ethical codes, on the other hand, are designed to specify ethical imperatives. Finally, law need not have any ethical content at all. In a strict sense, law is the dictate of the state; codes require ethical content as a matter of necessity.

If one accepts the legalistic argument, the codification of ethical precepts indeed becomes an impossible task, for it would have to specify how every precept applied in all possible combinations of circumstances. At this point, a code of ethics would be transformed into a statement of rules or laws. The penalties for violations of laws rest in legal sanctions, while the enforcement of ethical codes depends mostly on moral sanctions. Thus, to confuse ethics and law or what is legal for what is moral is a confusion of the first order and one common in Western society. In this regard, Senator William Fulbright has noted:

> One of the more disturbing aspects of this problem of moral conduct is the revelation that among so many influential people morality has become identified with legality. We are certainly in a tragic plight if the accepted standard by which we measure the integrity of a man in public life is that he keeps within the letter of the law.[15]

Members of the military must recognize the distinction between law and ethics, a distinction that has been central to ethical thought since the Greek city-state. A legal order may be clearly immoral as was affirmed in the Nuremberg trials and as in the judgment of General Yamashita, to say nothing of the more recent Calley case. The point is that a code of ethics is not a code of law; a code of ethics is stated far more generally than law, its applications cannot be specified as precisely as in law but require judgment, and its enforcement depends more on moral sanctions than on legal ones. To argue that a code of ethics for the military is impossible or useless because it does not meet the requirements of a code of law is to make a major error.

The implicit fear is, of course, that an ethical code would not be so specific as to allow for the assessment of legalistic penalties and protections. Or perhaps it would allow penalties to be assessed for transgression of ethical precepts in instances that would be far less clear than in an adversary procedure in a court of law. One officer at the Army War College reflected this attitude when he remarked that he had no objection to a code of ethics as long as it wasn't written down. He feared it would then become the basis for judging the behavior of officers within the profession! Both fears are well founded, but that is the nature of ethical choice. One cannot expect an ethical code to operate like a body of law, nor can one expect laws to operate like ethical codes.

6. Under certain circumstances, the propositions of a military code may conflict. Ethical precepts within the same code may indeed conflict, but the soldier must still choose among the obligations he has to observe when he cannot observe both. Ethical action involves the ability of a member of a profession to choose which of two obligations he will observe when circumstances prevent him from doing both. Thus, the fact that some of the precepts of a code may conflict in some circumstances is not an argument for failing to establish such a code in the first place. It is merely to describe the nature of ethical dilemmas in an empirical world.

Some people suggest that no single code, while it may establish ethical goals for a profession, can legislate the degree of value any member of the profession will attach to different and conflicting precepts. Accordingly, because individual members of the profession place greater value on some obligations than on others, the value of an ethical code is negated. Again, this position reflects a failure to understand the nature of ethical codes. Some members of the military profession, even in the same circumstances, may choose to observe different obligations when they cannot observe all relevant ones. Again, that is the very nature of ethical choice. The reason why a person observes one obligation over another when circumstances prohibit observing both is precisely because the person judges that one obligation in a certain set of circumstances ought to take precedence over another because it is more valuable. Nonetheless, in the code all postulates have equal value. It is the circumstances in which the precepts apply that require the individual soldier to judge that one is more valuable than another. But as precepts per se they are all equally ethically imperative insofar as they require their obligations be observed. They are also equally imperative in the sense that they can be raised by the principle of universality so that if all individuals or soldiers carried out the precepts we would still judge such action as ethical. Thus, the fact that one soldier may value one obligation over another in a different set of circumstances from another soldier does not negate the value of a code in stating the obligations that he must observe in the first place.

7. All codes are futile because they can be misapplied within the military community. The existence of a code does not guarantee compliance. This is a rather obvious objection and a minor one as well. No one seriously suggests that the mere promulgation of a code of ethics for the military will guarantee its observance. On the other hand, codes are statements of what men ought to do, and it is unlikely that one can establish ethical standards without codes, nor can one expect individuals to carry out their obligations until they know what their obligations are. From a negative perspective, one cannot correct or judge ethical practices without at least some idea of what the standards of ethical action are. A good code says nothing about bad practice. The promulgation of a code will not in itself produce ethical behavior; it will simply specify those obligations against which ethical behavior or the lack of it can be measured. Indeed, without such a standard, judgments about ethics become very difficult, especially when the profession attempts to render a communal judgment about the actions of one of its members.

8. The creation of a formal code of military ethics constitutes a danger to men of the profession who observe the code. Members of the military could come to perceive obedience to a code as relieving them of all obligations for moral choice by simply obeying the code. In this case, obedience would become a substitute for ethical judgment. Once again, this is an argument that suggests that the perversion of a code reduces the value of codes. Following the obligations expressed in a code of military ethics presupposes that the soldier knows *why* an obligation ought to be observed. Blind obedience to a code of ethics that is not understood is not ethical action at all; it is merely blind obedience. However, if in fact the code clearly does apply in the specific, then it will constitute an ethical imperative. Finally, obedience to a code does not remove ethical responsibility from the individual. Indeed, observing an ethical precept of a code without regard for the circumstances in which it applies can easily become an unethical act because the circumstances were such that the soldier should have followed another precept of the code. No military

man can ever escape ethical responsibility by simply following the precepts of a code, any more than he can escape ethical responsibility by following the precepts of the law or the orders of his superiors. Ethical action requires that individuals be aware of their obligations and know why their obligations ought to be observed. It also requires that observance of obligations be applied with an understanding of their impact, given the circumstances in which they apply.

9. While a code's purpose may be noble, a code that would counsel ethical action is needed only if one believes that men are bad and cannot be relied upon to do what they ought to do by themselves. Thus, a code of ethics is needed because men are essentially corrupt, and if men are essentially corrupt, then the mere provision of an ethical code will not correct this condition. Again, this is an argument that is misplaced; codes provide statements of obligations which men ought to live up to and which they must live up to if they are to be members of the military profession. By attempting to live up to a code, members of the profession are ennobled in the very striving. Yet, even corrupt men are not totally corrupt all the time, and ethical guidance as to what they ought to do seems to be the first step in getting them to do it. Codes are brought into existence not because men are corrupt, but because men are capable of great moral acts. It is only a secondary function of an ethical code to utilize the code to discern unethical acts; its primary purpose is to compel ethical acts.

Some of the arguments against the creation of a formal code of military ethics emerged succinctly in a speech made by the superintendent of the United States Military Academy at West Point in 1979:

> It may be, however, that we should make the point more sharply and strongly. Military service does require a certain basic pattern of commitment in ethical beliefs. But... it is not possible to prescribe in advance and in detail for every situation. And unthinking acceptance of a set of ethics prefabricated by others seems to us to have little promise for American military officers.[16]

This statement contains three objections to a code of military ethics. First, a code of military ethics would be unworkable because "it is not possible to prescribe in advance and in detail for every situation" the manner in which a code would apply. But there is no need to prescribe in advance and in detail the manner in which an ethical precept will apply in given circumstances. To make such a requirement is to confuse ethics with law. Ethics requires judgments about what obligations must be observed when they come into conflict. The idea that it is necessary to prescribe in advance for every situation in which an ethical precept will apply not only misinterprets the nature of ethics and confuses ethics with law, but it also reduces the officer to a bureaucrat. It removes from the officer the central responsibility for judgment in carrying out his obligations and seeks to substitute for it a rule book that can be used to decide what to do without making choices. In effect, the superintendent does not understand what ethical judgment is all about, namely, choosing what obligations ought to be observed under given circumstances.

The superintendent's implication is that because one cannot specify in detail how ethical obligations will apply in varying circumstances, one ought not to formally prescribe the obligations in the first place. However, if a formal code of military ethics is not developed, there is no way to teach ethics to begin with. There is only the hope that ethics will be "caught" from one's experiences. One might argue that good ethical standards are not taught as much as they are caught by being a member of a community and absorbing the informal lessons that emerge from the ambience of the military profession. In this view, the soldier acquires a set of obligations that are informal and unspoken but nonetheless clear to new members. The difficulty with this view is that it is an act of faith more than a statement of empirical conditions. There is very little evidence that the American military today possesses a set of core values that are clearly recognized, understood, and enforced by the profession itself. If we look at the difficulties that surfaced during Vietnam and continue to plague the military, exactly the opposite seems true.

The fact of the matter is that one abandons ethical responsibility for the profession if one simply assumes that sound ethical sense will be absorbed through a kind of social osmotic process. If ethics is to be taught, the soldier must be aware of his ethical obligations, and the only way to make him aware is for the profession to state them openly and formally. Precisely because learning ethics is difficult, ethical obligations must be clearly stated for the members of the profession.

The superintendent's third objection is that a "set of ethics prefabricated by others seems to have little promise for American military officers." The argument has some validity if one is talking about a code that is merely imposed through administrative fiat. An ethical code in which individuals do not understand why obligations are required, why they are moral imperatives, and why the obligations bind is not a code of ethics at all. It is merely a set of slogans. Ethical codes, by definition, require that the individual soldier know why his obligations obligate. Thus, the superintendent is correct when he suggests that to impose an ethical code without a sense of moral reasoning as to why the code binds is unlikely to work. But he is wrong when he suggests that ethics cannot be formulated and that the reasons why ethical imperatives and obligations ought to be observed cannot be taught to soldiers. In fact, soldiers *must* be taught why obligations ought to be observed or they will not be able to act ethically in any meaningful sense of the term.

Ethics and the Service Academies

The military academies have made a sincere, long, and serious attempt to come to grips with the problem of ethics in the profession, but the results have been disappointing.

An examination of some of the elements of the West Point honor code, for example, indicates the academies' confusion about ethics. (Moreover, the many cheating scandals in the various service academies over the years raise some legitimate doubts as to how well ethical instruction is succeeding at the academies at all.) The problem of socializing individuals to an ethical code which defines a good officer goes far beyond the capability of the academies themselves. In the first place, the honor code at West Point is too general to provide much guidance. The stipulation, for example, that no cadet will lie, cheat, or steal, nor tolerate anyone who does, offers no real help to a soldier in search of moral guidance. The West Point honor code is actually more representative of the morality of the college fraternity house in that it confuses character traits with ethics. In addition, no effort is made to tie the code to specific military situations to which it may apply beyond campus life. It tends to address those problems most likely to be encountered by the cadets *at the academy* rather than the problems they are likely to confront once they enter the larger military profession.

The academy's honor code, especially the nontoleration clause, turns all brother officers into potential informers. The prerequisite of any successful corporative institution such as an officer corps is to inculcate within the individual officer a sense of what he ought to do as an ethical soldier. This most certainly does not require that every member of the corps keep a constant watch on the actions of every other officer. Other service academies allow the individual some discretion and judgment in dealing with the unethical conduct of his peers. Their experience has been uniformly good, with the Air Academy providing an excellent example. When every comrade is a potential informer, the individual is forced to "play it safe" and to focus upon trivialities. More importantly, the tendency to focus on minor transgressions of rules and equating them with moral transgressions tends to destroy any sense of community based upon mutual trust, as well as a community attachment to a set of larger values. It creates officers who depend upon their ability to manipulate the rules of the system as a substitute for ethical judgment. When in doubt it is much easier and safer to follow the rules. Such lessons are appropriate for the officer who would be an entrepreneur; they are deadly for those who perceive themselves as members of a corporative professional institution.

Even the instructors at West Point openly acknowledge that the academy honor codes have very little relevance to conditions in the "real" profession. The degree to which precepts of the code can be transferred to behavior, especially as that behavior is supported by the norms of the profession at large, is minimal. A common occurrence is one in which a young academy graduate reports another officer for an honor offense. Such situations no doubt lend humor to daily life at the company level, but clearly imply that the officers in the "real army" do not expect the provisions of the honor code to apply to the profession at large. As part of his education, the young academy officer learns that the provisions of the honor code are not applicable or enforced within the profession. Consequently, after having these precepts drilled into him during his four years at the academy, after some of his classmates have been dismissed for failure to observe the code, and, perhaps, after having turned in a comrade for violating the code, the young officer is suddenly confronted with the fact that almost no one in the profession at large takes the code seriously. Inevitably, after a period of confusion, the search for another ethic will begin. Here the young officer finds only the informal, intuitively perceived, and experienced learned rules of the military entrepreneur. He finds no equivalent to the code of honor he learned at the academies for the most elementary reason: the military profession has never developed or promulgated such a code. From the point of view of professional ethical guidance, he is very alone.

The military profession's failure to develop an ethical code that can be applied to the soldiery as a whole may be traced to the practices in the academies. It is informative to note that the superintendent at West Point suggests that:

> The military academy therefore insists on certain institutional values chosen for their appropriateness in contributing to the self development of moral persons and for constituting standards to which all Army officers can proudly ascribe.... The military academy acknowledges that its graduates may and will subscribe to a variety and range of ethical systems and anticipates that all cadets will analyze closely and evaluate rationally their own moral beliefs. If graduate cadets are to serve as officers in the way they should, however, their own examined moral beliefs need to be in harmony with the values and standards requisite to an officer of the American Army.[17]

Paradoxically, while West Point agrees to the necessity of a statement of ethical principles to which officers must adhere, it refuses to develop a broader sense of ethics which the graduate officer would find applicable once he left the academy and entered the actual world of the military professional. Accordingly, several faculty members have pointed out that the academies do not seek to teach or inculcate prescriptive ethics. In its place, the academies are content to teach descriptive ethics and to acquaint the cadet with a history of ethical theories. It is expected that from this survey of ethical perspectives the cadet will extract those ethical precepts with which he feels most comfortable.

A similar approach is evident at the Naval Academy. In 1976, when Vice-Admiral William Mack was commandant, he became concerned about the problem of morality within the cadet corps, and so he instituted a new course entitled "The Officer and the Human Person." At that time, he stated succinctly the notion of descriptive ethics which formed the basis for his view of ethical training. He said of his cadets, "They were exposed to several sets of ethics and morals. They could take their own and keep them or accept others. I didn't care what kind as long as they had a system of their own."[18] The service academies generally share his point of view, ignoring the fact that it matters quite a lot which precepts are finally selected. The descriptive approach fails to develop and systematize a specific code of ethical action required of all officers and soldiers in the profession. Moreover, it makes no attempt to inculcate a code of ethics, the observance of which comes to define the "good" officer and separate him from the "bad" officer. The attempt is not made because it is assumed that the individual's sense of integrity will ultimately provide him with such a code by which to act as a soldier. It is also assumed that the ethical values which an individual brings with him when he enters the academy are sufficient for the military profession at large and that they require no further codification or inculcation. The final assumption is that a group of officers, each secure in his own sense of personal ethics, will produce a code that is appropriate for the soldiery as a professional community. This notion is a curious application of the free market idea of ethical egoism to the study of ethics, and it is simply incorrect.

The ethical failures of the military profession have only partially been the result of the failures of the individual officer's sense of ethics, and this is often true of the cadets at the academies as well. The profession as an institution has failed to develop a sense of ethics that is supportive of individual notions of integrity and to ensure the training of soldiers in what the ethical standards of the profession are. It is naively incorrect to assume that a collectivity of individual officers with individually derived sets of ethics will automatically produce an ethical code for the professional community as a whole. Precisely the reverse is likely to be true. A community code of ethics can be used to socialize novices to the profession, and this process of socialization helps clarify the "price of belonging" which is so necessary to the ethical cohesion that should be characteristic of a professional institution.

The idea that ethics is somehow different at the academies than it is in the profession at large represents a curious kind of ethical egoism in itself. Louis Sorley has pointed out that "the principal justification for maintaining a national military academy has been to provide a nucleus of officers who can leaven and set standards for the entire officer corps in terms of dedication and ethical standards."[19] The service academies widely accept this "missionary approach." The missionary approach assumes that somehow academy graduates will be instructed in codes of honor and ethics that are different from and far above those found in the profession at large. These academy graduates will then go forth and act as missionaries; by their own example and honor, they will engender within the profession a sense of ethical behavior that will lift the rest of the profession to these high standards.

This approach to ethical development reflects a sense of egoism and arrogance that is unworthy of academy graduates. It is also insulting to the rest of the brotherhood of arms. Is there any reason to believe, for example, that the men who come to the profession from ROTC, from other military colleges, from the Officer Candidate School (OCS) ranks, or even by direct commissioning are in any sense less dedicated, less talented, less ethical, less willing to bear the burdens of military service, and, indeed, less willing to bear the ultimate sacrifice of their own lives than are academy graduates? There is no good reason to assume that academy graduates have either greater potential for ethical action as cadets or after they leave the academy than any other officer. To suggest so is a slur on the rest of the profession.

Even if one could create a corps of military missionaries at the academies, it would still not solve the problem of ethics for the profession itself. In the first instance, it would leave out 90 percent of the officers of the profession who are not academy graduates. More to the point, as long as the academies themselves do not have a formalized code of ethics to which officers in the rest of the profession can adhere, even moral missionaries who act in support of their own internalized codes are likely to be perceived as merely bizarre. Without a stated code of ethics which all within the profession share, even the ethical behavior of the best intentioned officer may be misunderstood. Without a stated code of ethics for the academies which is congruent with the ethics of the profession as a whole, to continue to teach descriptive ethics and to justify it in terms of some vague "missionary effect" is to miss the point of ethical instruction, which is to inculcate a set of obligations that individuals must live up to and an understanding of why those obligations ought to be observed.

The military academies do not prepare the young officer for the ethical choices and situations he may have to confront within the profession. The major fault of the honor codes is not that they set impossible standards or even that they are outdated, but that they are largely irrelevant and tend to substitute managerial for corporative values. Consequently, the military has come to expect that a group of officers of high personal integrity will somehow come to constitute an ethical community with a well-defined sense of professional ethics as well as the courage to act upon them. It represents an attempt to transfer the notions of ethical egoism and entrepreneurialism to the military profession under the guise of ethics by allowing the forces of the ethical marketplace to operate unchecked. The attempt is as absurd as the economic analogy upon which it is premised, and, above all, it has failed. What is required is a code of ethics for the military profession that would become the basis for socializing young novices to its membership.

The Ethical Dilemma

The experience of the military profession over the last two decades has produced a military careerism so exaggerated that the protection and advancement of one's career has become the highest operant value for a large number of officers. This change in traditional military values has not been without its practical effects on the profession's operational capabilities. In general, however, this change has produced a series of ethical failures, with soldiers initiating or participating in unethical policies and actions as a means of advancing their careers.

The exaggerated emphasis upon careerism as the ultimate value can only occur in a military profession that has failed to develop strong formal ethical doctrines. Accordingly, in their place such shorthand injections as "it all counts for twenty," "don't rock the boat," "get on board," and "don't fight the problem," while often failing from an ethical perspective, actually become valuable pragmatic guidelines for career success. Clearly, resistance, dissent, and protest as ethical alternatives to acquiescing in actions with which soldiers seriously disagree are not strongly established practices in the profession.

The marked failure of officers to retire in protest, to resign, or to dissent openly from policies they feel to be morally repugnant or not in the best interests of their commands or country is a symptom of a much deeper malaise. The Vietnam War did not create the propensity "to go along" or to become a "team player" in support of questionable policies. Rather, that propensity was established as a consequence of the shift away from traditional, corporative, and ethical institutional values toward entrepreneurial and managerial ones. Once the military began to be regarded as an occupation rather than a vocation, soldiers began to assess the responsibility to obey or object to actions largely in terms of the requirements of the "ethical marketplace" in which their careers were being played out. If an army remains an ethical institution, the soldier's responsibility to object to or to support questionable policies can be assessed against the more secure grounds of a sense of shared ethical values. The tendency to support policies about which a soldier may have serious ethical misgivings is a normal consequence of the shift within the profession away from clear ethical communal values toward entrepreneurial ones and the attendant careerism that results. In short, nothing in the values of the profession itself can shield the soldier from the pull of exaggerated self-interest.

What is at stake is the degree to which hostile values within the profession erode the necessity to develop codes of professionally based morality. No one is suggesting that individual soldiers are not men of integrity in their *personal* lives. What is at question is the degree to which a developed sense of *professional* morality can be relied upon to support the decisions of soldiers which are based in their concepts of what constitutes personal integrity. Is there a tension between the ethics of the individual soldier and that espoused by the profession in which he claims special membership? In a true community of professionals, there is a strong convergence of individual and professional ethics.

Few individuals can be expected to stand alone for very long against the dictates of a powerful organization sustained only by the comfort that what they do is supported by their personal sense of what is right. It is, therefore, unrealistic to expect soldiers to question orders, resist their execution, dissent from superiors' views, and, in the extreme, even refuse to comply with orders and policies they find morally objectionable unless there is some reasonable expectation that what they are doing is supported by the formalized ethical precepts of the profession in which they act. As things now stand, the erosion of professional corporative values may be so complete and their replacement with the entrepreneur's values so thorough that no extant code of ethics could define ethical action for a member of the profession. Worse, there is no place where a soldier can go to learn such a code. To the extent that membership in a profession requires observance of a code of professional ethics, it is imperative that the code be evident to all members of the brotherhood. If not, the profession will have no ethical mechanism for socializing its own members, especially those newly commissioned officers, to a higher ethical standard that defines what membership in the profession entails and that serves as a guide to ethical judgment on and off the battlefield.

It might be objected once again that there is no need for a formal ethical code, that the informal assents given to informal norms serve to provide a mechanism for socializing young novices to the profession. In place of a formal code, the soldier is encouraged to rely on what he can deduce informally through his experience. The difficulty is that the ethics of the marketplace inevitably becomes the basis for these informal norms. In the absence of formal competing values, self-interest is allowed free rein in the ethical judgments of the soldier. When these conditions obtain, the values attendant to a corporate ethical system are already dead. Careerism as the highest value is of necessity an individual reward when the true measure of a professionally oriented ethical soldiery, especially its officer corps, is the extent to which individual actions serve communal goals. While informal norms are important, they are unlikely to be adequate guides for ethical action because they come from an environment characterized by entrepreneurial self-interest. Precisely because they are informal, one suspects that such norms are not derived from some higher notion of the good officer, namely, his ability and willingness to abide by a code of ethics congruent with his own sense of ethical integrity and that of the profession. This descriptively ethical approach is a contradiction in that no attempt is made to formalize the rules of ethical behavior by which a soldier's actions may be judged.

A Code of Military Ethics

A first step in the attempt to inculcate ethics is to develop an ethical code for the officer corps and soldiery that would be the basis for socializing its young novices to the profession. An organizational milieu that lacks a sense of community ethics must inevitably result in little or no professional support for ethical decisions, thus producing irresistible tensions between the press of conscience and the attractions of career. Only a few, very strong soldiers can be expected to withstand this pressure without the support of the ethical ambience of the profession itself. In the absence of overt professional support, most soldiers abandon personal responsibility and acquiesce in the norms of the organization. The failure to abandon the ethics of the entrepreneur and to develop a code of ethics congruent with the nature of the military profession leaves its members without adequate ethical guidance in the performance of their professional duties. Hence, the soldier as entrepreneur becomes the dominant model for young novices because he appears most likely to prosper in the environment in which he must operate.

The code offered here cannot stand alone; it is only one variable in a highly complex equation of insitutional change. The most that can be said for it is that it seeks to formulate some basic ethical precepts for the profession and to alter the background ambience so vital to the successful inculcation of ethics in a profession. It does not attempt to state the sum of the soldier's obligations, but only those obligations that are most directly relevant to his role as a professional. The code has the singular task of seeking to detail the soldier's obligations against the general background of a profession dedicated to values associated with *combat, command,* and the *responsibilities* and *obligations* derived from them by a profession whose primary task involves the risk and taking of human lives in large numbers. The military, as described earlier, differs from other occupations and professions, and the center of that difference is the unlimited liability of its members. Thus, ''we should agree at the outset to take as our ideal the officer successfully performing his duties in the highest test of his profession—leadership of men in combat.''[20] The values of the staff officer, the organization man, the ticket puncher, the military manager, and the entrepreneur are rejected or subordinated to values associated with combat risk and death.

The purpose of the code presented here is to develop and sustain the values, habits, and practices traditionally associated with the military as a special community of brothers who share special values and responsibilities. It seeks to define the soldier in a manner that is far more extensive and intensive in terms of his values and responsibilities than we have recently come to associate with the soldier as manager or entrepreneur in pursuit of a successful career. The purpose of any code goes beyond an attempt to establish a sense of community. It also seeks to engender the feeling of belonging to a special profession by prescribing and proscribing certain values and actions which, in themselves, come to define legitimate membership in the professional community. In this context, the profession will become and remain ethical only when its members recognize and observe the prescriptions of the code that define it as special.

The Soldier's Code of Ethics

The nature of command and military service is a moral charge that places each soldier at the center of unavoidable ethical responsibility.

A soldier's sense of ethical integrity is at the center of his effectiveness as a soldier and a leader. Violating one's ethical sense of honor is never justified even at a cost to one's career.

Every soldier holds a special position of trust and responsibility. No soldier will ever violate that trust or avoid his responsibility by any of his actions, no matter the personal cost.

In faithfully executing the lawful orders of his superiors, a soldier's loyalty is to the welfare of his men and mission. While striving to carry out his mission, he will never allow his men to be misused in any way.

A soldier will never require his men to endure hardships or suffer dangers to which he is unwilling to expose himself. Every soldier must openly share the burden of risk and sacrifice to which his fellow soldiers are exposed.

A soldier is first and foremost a leader of men. He must lead his men by example and personal actions; he must always set the standard for personal bravery, courage, and leadership.

A soldier will never execute an order he regards to be morally wrong, and he will report all such orders, policies, or actions of which he is aware to appropriate authorities.

No soldier will ever willfully conceal any act of his superiors, subordinates, or peers that violates his sense of ethics. A soldier cannot avoid ethical judgments and must assume responsibility for them.

No soldier will punish, allow the punishment of, or in any way harm or discriminate against a subordinate or peer for telling the truth about any matter.

All soldiers are responsibile for the actions of their comrades in arms. The unethical and dishonorable acts of one diminish us all. The honor of the military profession and military service is maintained by the acts of its members, and these actions must always be above reproach.

The nature of command and military service is a moral charge that places each soldier at the center of unavoidable ethical responsibility.

The code affirms that command is a "moral charge" that places the soldier at the center of ethical responsibility and a matrix of professional obligations. Accordingly, command is never just one more "ticket to be punched" on the way to making the next promotion. Command is the very essence of military life; it has its prerogatives, but mostly it levies awesome responsibilities for the lives of one's men which come to define a crucial aspect of the way of the soldier. It will avail a military organization nothing if its staff work and support are all in order, but it finds itself incapable of producing good field commanders. War is the art of conflict, and command is the final expression of that art. Command *responsibilities* assume an almost mystical place in the litany of military values and occupy a central place on the altar of military ethics. The failure to recognize command as the center of the profession's ethical responsibility and to recognize that the soldier must bear this responsibility ethically is to deny that there is any difference between the profession of arms and any other occupation.

A soldier's sense of ethical integrity is at the center of his effectiveness as a soldier and a leader. Violating one's ethical sense of honor is never justified even at a cost to one's career.

The notion that ethical integrity is at the center of a leader's effectiveness suggests that the compromise of moral standards can never be truly hidden from brother officers or, indeed, in most instances from the men in his command. The concept is simple enough: some things are not done. There is a line beyond which a truly ethical man will not go, and in the profession of arms that line is often drawn very clearly. The notion that one has to "go along to get along" is rejected as the first step down a path leading to ever greater compromises of ethical standards. Ethical men cannot compromise their ethics and remain either moral men or good soldiers.

Every soldier holds a special position of trust and responsibility. No soldier will ever violate that trust or avoid his responsibility by any of his actions, no matter the personal cost.

The code maintains that to be a soldier is to occupy a "special position of trust" and that this trust—which goes hand in glove with the virtue of integrity—must never be violated regardless of the consequences to the soldier's career. If there are circumstances in which it must be violated, career advancement is never among these. There will almost certainly come a time in the life of every soldier when he will have to choose between pleasing his superiors or being true to himself and to the values of the profession. At times the rewards for betraying personal values as well as professional values are great, too often reflected in promotion or other awards. The code sets up the simple standard that the soldier's personal integrity and position in the military are inseparable; to violate one is to violate the other. A soldier must never betray his position, for to do so is to betray his men, his profession, and himself.

It may be objected that this aspect of the code is too individualistic and that it emphasizes the role of individual conscience over obedience to authority. In response, one should point out that only individuals are truly capable of ethical action, not social fictions. To say that the "profession" did this or the "organization" did this is, in a strict sense, to speak in fictions. Ultimately, only individuals are capable of ethical actions and only individuals can be held responsible for the consequences of their acts. The focus of ethical decision-making must remain on the individual soldier. From this perspective, the code is not too individualistic and does not stress individual conscience at the expense of authority. It merely recognizes that a soldier acting within an organizational capacity or within the confines of a profession still cannot abandon his ethical responsibilities, nor, in fact, can he abandon his conscience. All ethical action remains individual action and no less so because it occurs within the context of the military profession.

> In faithfully executing the lawful orders of his superiors, a soldier's loyalty is to the welfare of his men and mission. While striving to carry out his mission, he will never allow his men to be misused in any way.

The proposition that a soldier's loyalty is to the welfare of his men should not be interpreted to mean that he should be fearful of putting them in harm's way. Indeed not, for it is the very essence of the military profession to engage in combat. It may also appear that there is a tension between the precept that an officer's loyalty is to the welfare of his men when the emphasis should really be on loyalty to his superiors and mission. Loyalty to one's men has relevance in the context of loyalty to the mission which must be accomplished in a particular set of circumstances. The point is that in the event of a perceived conflict between looking after the welfare of one's men and following the orders of an incompetent superior, the commander must have another moral anchor besides the absolutist dictum to obey orders. The tension is, therefore, deliberate in order to force a moral choice. The idea that a commander's first loyalty is to the welfare of his men in no way implies that he should not be prepared to carry out his mission. It only implies that in carrying out the mission he is responsible for the consequences and that he remains the focus of ethical responsibility. As General Matthew Ridgeway pointed out, "a commander has as deep a duty to the men with whose lives he is temporarily entrusted as they have to him—and part of that duty is to see that those lives are not needlessly squandered."

A commander must never allow his men to be squandered or used in a manner that is not directly related to the true purpose of command. Danger is central to combat; the object is to expose one's men to that danger only in pursuit of legitimate military objectives and goals. What constitutes a legitimate objective and what constitutes an illegitimate one is a determination that each commander may have to make when he finds himself in a situation that makes him doubt the wisdom of his superiors in ordering certain actions. The crucial point is that he must be prepared, in defense of his oath and his ethics, to question his superiors and, if he deems it necessary, even overtly refuse to expose his men to risk to achieve doubtful objectives. He must never allow himself or his men to become mere tools for the advancement of his superiors.

A soldier will never require his men to endure hardships or suffer dangers to which he is unwilling to expose himself. Every soldier must openly share the burden of risk and sacrifice to which his fellow soldiers are exposed.

A soldier is first and foremost a leader of men. He must lead his men by example and personal actions; he must always set the standard for personal bravery, courage, and leadership.

A good soldier must share the risks and dangers of combat to which his men are exposed and be willing to suffer the ultimate sacrifice of his life if conditions warrant. This, above all, is what differentiates the military professions from all other occupations and even other professions. The basic lesson which commanders since the time of Thermopylae have always known is that to be effective a commander must be seen on the field of battle with his men. A good soldier realizes that men will follow his judgment if they are convinced by his example that he has as much to lose as they have and that he is prepared to risk his life in their defense. If a military force is to be effective on the field of battle, every soldier must come to accept the responsibility of exposing himself to the dangers to which his men are exposed and, if necessary, to fulfill his obligation of unlimited liability along with them.

A soldier will never execute an order he regards to be morally wrong, and he will report all such orders, policies, or actions of which he is aware to appropriate authorities.

One of the stronger precepts of the code is that which affirms that a soldier cannot escape responsibility for his actions before the law, his peers, the profession, or his own conscience. The problem of "following orders" is faced day in and day out by numerous officers outside any zone of combat. For example, the system of checking readiness status in use today forces hundreds of officers to report equipment ready for action when in fact they know it is not. In other instances, soldiers are witness to actions which they know to be wrong or dishonest and which they conveniently ignore on the grounds that to report these violations will result in a poor efficiency report, the denial of a promotion, or some other punitive action. Accordingly, the injunction that an officer or soldier will never execute any order he regards to be morally wrong goes far beyond the bounds of action that one is likely to encounter in a combat zone. It also addresses the kind of activities that often permeate a highly bureaucratic military organization. The directed moral charge is that a soldier is in violation of his honor—in betrayal of his ethical trust and special obligations—if he has knowledge of unethical occurrences and either acquiesces in them or fails to report them.

It might be argued that any precept that requires a soldier never to execute an order he regards to be ethically wrong invites disobedience. This is not the case. Such a precept merely restates the commonly agreed upon ethical principle that no individual can ever escape responsibility for his actions and that following orders is never an excuse for acting unethically. The object of ethical training is, of course, to train soldiers to recognize ethical dilemmas and to work their way out of them by making choices among conflicting obligations. Nonetheless, the responsibility remains with the soldier, and no attempt is made to invite disobedience by suggesting that the soldier ought not to carry out orders that are not unethical. The precept only points out that there are likely to arise instances in which a soldier perceives grave moral tensions and conflict between what he is ordered to do and what he feels is morally permissible. In such instances, it is simply no moral or legal defense or excuse to abandon one's ethical responsibility and obey orders. Again, we return to a fundamental ethical principle—that only human beings are capable of ethical action, and they must remain the focus of ethical responsibility, accepting responsibility for the consequences of their acts.

No soldier will ever willfully conceal any act of his superiors, subordinates, or peers that violates his sense of ethics. A soldier cannot avoid ethical judgments and must assume responsibility for them.

Among the most notorious and corrosive norms which the military borrowed from the business world are those of the "team player" and of "getting on board." These norms require of the military professional, as they do of the business executive, that as subordinates they be "loyal" to their superiors and that certain policies be followed unquestionably because to challenge them places the individual in a position of appearing to "fight the problem" and not to trust his superiors. Subordinates are expected to be "part of the team," to be loyal to their superiors. But what does it mean for a subordinate to be loyal? Surely, whatever loyalty is owed to a superior cannot ethically be construed to include a willingness to conceal his failings and shortcomings, especially when they have a bearing upon a unit's ability to perform its mission or when men may be killed. There must be a higher loyalty, and that is to one's men, the profession, and its code of ethics. Loyalty can never be interpreted to mean that policies, orders, or actions which the soldier thinks are detrimental to the success and effectiveness of the unit are to be tolerated by subordinates. Loyalty to one's superiors is never anything but a conditional relationship predicated upon the continuing perception that one's superiors are acting ethically in the conduct of their professional responsibilities. Once superiors begin to act otherwise, even that relationship is dissolved, and the obligation to make appropriate authorities aware of existing circumstances takes precedence.

No soldier will punish, allow the punishment of, or in any way discriminate against a subordinate or peer for telling the truth about any matter.

No profession can respond to the problems that may develop within it, especially the problem of unethical behavior, unless it has adequate information upon which to take corrective action. There is an organizational as well as an ethical imperative for telling the truth to one's superiors, peers, and others who occupy posts of authority within the profession. Members of a profession or any organization will feel at liberty to tell the truth only if they can be reasonably certain that superiors and peers will not regard telling the truth as "fighting the problem," as not being a "team player," and, in some instances, as disloyalty bordering on treason. The profession must never strike at any member for exposing even the most heinous of crimes or any shortcoming within the profession itself. The far greater crime would be to hide the truth, not only because to do so is unethical, but also because it can have devastating practical consequences on military capabilities and operations. It may well be an open question whether the truth will make one free in any given instance, but clearly falsehoods never contribute to freedom. If the soldier is made responsible for a higher ethical code in the same way as a member of a monastic community is so responsible, there can be no justifiable punishment for revealing the truth. To punish the truthsayer negates the value of other ethical precepts.

All soldiers are responsible for the actions of their comrades in arms. The unethical and dishonorable acts of one diminish us all. The honor of the military profession and military service is maintained by the acts of its members, and these actions must always be above reproach.

Finally, we must never lose sight of the fact that the military profession is, in many respects, akin to a religious brotherhood, or at least ought to be. All soldiers are brothers in the same sense that monks are brothers. In specific terms, this means that the dishonorable actions of one soldier necessarily besmirch and diminish all. Responsibility becomes collective in that the community of brothers is unwilling to tolerate or permit in their midst one who fails to observe the community's common ethical code and standards. Every soldier is thus responsible for his brothers, which imposes the awesome responsibility of ensuring that one's fellow soldiers remain true to the profession's ethical values. The profession must, therefore, be prepared to enforce its standards on the membership and to summarily dismiss those whose actions are "unbecoming" to their station, and testify to their inability or unwillingness to pay the price of remaining in the community. This responsibility falls most heavily upon the senior grades who occupy strategic positions within the profession, for they above all others have the power to enforce the code. Indeed, a basic precondition for the acceptance of any new values within a professional institution is the readiness of institutional elites to provide full and overt support to the new values. So it is with a code of ethics. Unwillingness to remove the "deadwood," or any hint that a "protective association" may be operating, or further evidence of hypocrisy at the top will diminish the compelling power of any ethical code. With rank goes power, and with that power goes the responsibility of ensuring that the ethical code of the military profession is enforced at all levels of command.

Conclusions

Even the most trusting students of military organization know that the mere promulgation of an ethical code for the military will not ensure its effectiveness. Yet, one cannot resist pointing out that it is difficult to expect the profession and its members to act ethically in the absence of a code. The suggestions made here might be found to be unworkable, but surely the proposition is acceptable that there is at least the possibility of developing a code of working ethics for the profession of arms. The time is long past when we can continue to rely upon the ethics of the marketplace to provide standards of behavior for the military profession. Military units that have no ethics will fail to develop a sense of community and will then fail to develop strong personal ties among their members. Without these bonds, combat effectiveness as a function of cohesion rapidly falls off. In this very practical sense, if for nothing else, military ethics is necessary to the profession.

In a broadly human sense, military ethics is the only thing that can make the horrors of war bearable. Without the psychological fortification of being a part of a larger community and brotherhood in his military unit, the soldier can easily fall into the trap of Eichmannism—the value-free technician committing all manner of horrors because he was "just following orders." Or he will become an entrepreneur and "destroy the village in order to save it." Without the binding tie and sense of comfort that military ethics engenders, the soldier confronted with a moral dilemma will find himself very alone.

War, after all, is a *human* activity. If it will not go away, the task is to limit its individual and collective horrors. In this sense, military ethics supports a sense of community to form the basis of a profession that gives a human and humane dimension to the special task of war with its awesome and special responsibilities. We dare not fight without it, for without it we may yet win the battle, but the fruits of victory will become ashes in our mouths, and we may cease to be men.

Notes

1. Maxwell D. Taylor, "A Professional Ethic," *Army* (May 1978): 18.
2. Ibid.
3. William M. Lynn, "The Military Profession: What Is It? *Army* (September 1971): 23-27.
4. Lewis Sorley, "Duty, Honor, Country: Practice and Precept," *American Behavioral Scientist* 19, No. 5 (May-June 1976): 638.
5. Ibid.
6. Taylor, "The Military Profession," p. 19.
7. Ibid.
8. Ibid., p. 21.
9. Ibid.
10. I am deeply indebted to Major Steve Brodsky of the Canadian Armed Forces and Royal Roads Military College for the help in generating many of the arguments against a code of formal ethics.
11. Steven Brenner and Earl Molander, "Is the Ethics of Business Changing?," *Harvard Business Review* (January/February 1977): 66.
12. Malham M. Wakin, "The Ethics of Leadership," *American Behavioral Scientist* 19, No. 5 (May-June 1976): 573.
13. Sorley, "Duty, Honor, Country," p. 643.
14. Ibid.
15. Speech by Senator Fulbright before the U.S. Senate on January 28, 1967.
16. From a speech by General Andrew J. Goodpaster before the Association of American Colleges, entitled "Moral Choices: Ethics and Values in the 80's," in Washington, D.C., February 4, 1979.

17. Extracted from the "Concept for Furthering Cadet Moral Development," which is the central planning guide issued by the USMA superintendent on the subject. The copy which I have in my possession is for academic years 1978-1980.

18. William P. Mack, "The Need for Dissent," *New York Times Magazine* (January 12, 1976): 25.

19. Sorley, "Duty, Honor, Country," p. 631.

20. Taylor, "A Professional Ethic," p. 20.

6 THE CHARACTER OF THE SOLDIER

One of the major purposes of a code of military ethics, aside from ennobling the profession and specifying its obligations, is to establish points of reference that can be used for the character development of the soldier. General Taylor is quite correct when he states that a code of ethics will aid the processes within the military itself which contribute to the building of good character habits among the soldiery.[1] Any treatise on military ethics must necessarily discuss some of the qualities of personal character that would be expected from the soldiery in general and the officer corps in particular.

Ethics and Virtue

In a society such as ours which has been hesitant to set ethical standards in clear, concise, and formal ways the tendency is to suggest that one can avoid the need for ethical codes by encouraging the development of certain character traits or virtues among members of the military. The assumption is that soldiers who manifest certain character traits will automatically transfer these traits into behavior. While character-building of the individual soldier is to be commended, it in no way guarantees that he will act ethically. In a sense, it is true that virtue applies to a definition of what an individual is, that is, what he is in being, whereas ethics has more to do with how an individual acts under certain conditions. Thus, ethics and virtue are related, but they are also conceptually distinct.

To understand the role of military virtues in building the character of the soldier requires knowledge of the distinction between ethics and virtue (i.e., character). Character encompasses an individual's personal qualities as a human being. In the military, for example, a good officer should be loyal, obedient, honest, truthful, and courageous. These qualities are not only good in themselves but are also useful to any man, especially the soldier. *But virtues or character traits are not ethics.* Virtues are not innate but must be acquired by teaching and practice. They are also traits of character rather than traits of personality, and they are stable and not simply transitory feelings that a person may acquire at certain times. Virtues involve a disposition toward certain actions, but they are not the equivalent of the actions themselves. Nor are they skills or abilities. Virtues are predispositions to act in some ways and not in others.

Thus, those who maintain that the acquisition of certain character traits by members of the military will ensure ethical behavior are not correct, for morality is not merely cultivating character. Ethics implies ethical action, not just the predisposition to act ethically. As one commentator has noted, "I am inclined to think that principles without traits are impotent and traits without principles are blind."[2] In order to understand the tension between ethics and virtue, one must recognize that virtues are a set of predispositions to act in certain ways which men tend to affirm as good, all other things equal. Ethics, on the other hand, are precepts about how men ought to act; they specify what men ought to do if they are to judged as ethical. One might then distinguish between an *ethics of virtue* and an *ethics of duty*. An ethics of virtue refers to those traits of character that are regarded as creating predispositions toward ethical action, whereas an ethics of duty has to do with the way one acts in terms of observing specified obligations. An ethics of duty stipulates precepts as to how one ought to act, while the attendant sanctions for motivating the individual to act are not totally external to the actor. The role of virtue in an ethic of duty is not to tell one what he ought to do (for the ethic of duty specifies that), but to ensure that one does it willingly and conscientiously by engendering a predisposition to act ethically, a predisposition that is conceptually and empirically distinct from the act itself. Virtues are *ways of being* rather than *ways of doing*, although they are inevitably connected when ethical action must be undertaken or ethical choices made.

Will the individual who has certain desirable character traits always act ethically? He probably will not. Virtues do not reveal what a person ought to do; rather, they tell what a person ought to be in a generic sense. From this perspective, one can examine the SS officer, the terrorist, the assassin, or any other such individuals who share certain characteristics with the military professional, namely, loyalty, dedication, self-sacrifice, courage, and self-righteousness. No one would rightly argue that the SS officer who possessed all these traits of character and yet carried out the massacre at Lidice or garrisoned the extermination camps was acting ethically. Nor would anyone suggest that the terrorist who possessed similar personal qualities was acting ethically as he planted a bomb in a crowded theater. Yet, if soldiers do not have the traits of character commonly associated with ethical action, they cannot reasonably be expected to act ethically at all, for virtues are predispositions to act. On the other hand, the mere presence of a number of virtues does not guarantee that the soldier will make ethical choices. Without some virtue ethical action is impossible, but it must still be recognized that the inculcation of character traits in and of itself will not produce ethical soldiers. The paradox is that men of great character are quite capable of committing grievously immoral acts.

Both the ethics of virtue and the teaching of virtue are necessary in the soldier's development into a human being whose sense of ethics and manner of being are in concert. No one, however, is born with a sense of virtue. Virtues, like ethics, must be taught, and this teaching most often occurs within social and organizational settings. Character traits facilitate ethical action and are a legitimate concern of the military profession in deciding who is allowed to gain membership and to remain in good standing. On the other hand, military academies, colleges, and staff schools have tended to equate the possession of certain character traits with ethics, ethical training, and even ethical action. This is a significant error. If in its training of the soldier, the profession focuses on the inculcation of character without any consideration for ethical codes, the effort is likely to be self-defeating. Wrong, too, are those who suggest that if only those men of highest virtue are selected from the populace, they will necessarily be men of the finest ethics. It is both a falsehood and a non sequitur to assume that virtue equals ethics. Finally, the military profession cannot abandon its responsibility to establish standards of character by arguing that those who enter the profession already bring with them a developed sense of virtue and that, accordingly, the military must accept individuals pretty much as it finds them. Such a suggestion is to imply that the profession cannot teach ethics or that it cannot or need not teach virtue. It is also to deny that the profession can ever separate itself in any meaningful way from the larger society by its level of ethical action or the character of its members. Thus, character development remains an important part of the profession's responsibility toward its members.

Any work on military ethics would be incomplete without a consideration of the virtues that have historically been associated with the character of a good soldier. This discussion addresses a considerable number of such virtues, giving equal space to most of them, although some are indeed more desirable than others. The order in which they are treated does not imply the order in which they ought to be taught within any system of military education.

In a discussion of virtue especially as it applies to the military, the notion of *perfect virtue* which is so common in philosophical treatments of the Greeks and among the scholars of the Middle Ages seems to this writer patently false. Perfect virtue is normally defined as a state of being in which the predisposition to act in a certain way results almost automatically from the presence of the virtue itself. Such a concept borders too closely on the concept of habit. Virtues are states of being, and they imply that a soldier ought to be a certain kind of person. But no human being can attain, in any meaningful sense, a state of virtue that is perfect. To affirm that he can may be a matter of philosophical conceptualization, but it flies in the face of empirical fact. If ethics requires that ought implies can, there is every reason to apply the same standard to the development of virtue. Men being what they are they will sometimes succeed at virtue and sometimes fail. So, too, in the inculcation of virtue; sometimes they will be virtuous and sometimes they will not. Goethe made this point when he said, "if you treat a man as he is he will remain as he is; if you treat him as if he were what he could be he will become what he could be." In short, virtues can be stated as an idealized sense of what the character of a good soldier ought to be. It is not in the possession of virtues *in the perfect sense* that one becomes virtuous and capable of ethical actions as much as in the *attempt* to develop those virtues in the perfect sense. There is almost a Platonic sense in which virtues can be seen as comprising a set of idealized predispositions of the human being, which in the striving to attain them the individual becomes ennobled and virtuous, although he never attains them in their ideal state.

If ethics is more than the conformity to internalized rules and also requires that one observe obligations that are externally imposed, it is clear that conformity to an ethical code must rest on an understanding of the reasons why the obligations bind. It also implies that one must be a certain kind of person, that one possess certain traits of character. Ethical choice and character are separate but intertwined qualities of ethical action. Those who would suggest that ethical codes will suffice in the absence of character virtues are bound to be disappointed, as are those who maintain that the inculcation of character traits alone is sufficient to produce an ethical soldier. What is required is an understanding of virtues and of how they relate to ethics. Let us now examine those military virtues which soldiers in the Western world have identified over the last two thousand years as being vital qualities of character.

JUDGMENT AND INTEGRITY

Judgment and integrity are more important than other virtues insofar as they integrate other aspects of the character of the soldier. Judgment is more concerned with ethical action than is integrity, which seems more related to moral being. Judgment integrates an ethics of duty, while integrity welds together an ethics of virtue. Without judgment or integrity, no ethical code would have much chance of being effective. At the same time, without a sense of integrity any sense of being a man of ethical character apart from ethical action could hardly develop.

Judgment is defined by the Oxford English Dictionary as "the action of mentally apprehending the relationship between two objects of thought; predication as in the mind; the critical faculty in the formation of personal or individual opinion as opposed to acceptance of doctrine or authority." In a word, soldiers must be able to judge not only what obligations take precedence over others under certain circumstances, but also why one obligation takes precedence. Judgment cannot be taught without allowing some attempt at its activity. A man can only be what he is becoming and, as Colonel Mike Malone of the U.S. Army War College is fond of saying, the best way to give a soldier the opportunity to be ethical is to give him the opportunity to behave unethically and watch him choose. Thus, a soldier who has judgment must be able to discern the connections between events as they occur in the empirical world and to discern the connections between his actions and their consequences; and, of course, he must be able to choose among them. But the act of choice follows upon one's ability to develop a "discernment of the mind" as to what courses of action are open. The quality of judgment is, therefore, central to observing any ethical code since judgment rests at the center of ethical action. Because judgment is the means by which a soldier chooses among obligations, it is a primary military virtue that should be developed to the extent possible within every soldier. Without it, ethical action is either not possible or gravely endangered in its manifestations.

Equally important as a military virtue is integrity. Integrity is derived from the Latin word *integer* meaning wholeness, entireness, or completeness. Integrity "is the condition of having no part or element wanting; the soundness of moral principles; the character of uncorrupted virtue especially in relation to truth or falsity." It provides an overall perspective as to where other virtues fit in an individual's overall character. Without a sense of integrity, without a sense that man must be an integrated whole if he is to be an ethical man, the teaching of ethics and virtue is likely to be pointless. The importance of integrity to character development in the military is clear enough. As General John Ryan, U.S. Air Force, has stated: "Integrity is the most important responsibility of command. Commanders are dependent on the integrity of those reporting to them in every decision they make. Integrity can be ordered but it can only be achieved by encouragement and example."[3] Any attempt to develop character virtues by the military must inevitably aim at developing the whole man in the humanistic sense that the soldier must realize that what he does within the profession is only one facet of his whole ethical being. He must understand that all these facets are bound together and that at times there will be conflicts among them. But the integral man understands that there is no escape in a false compartmentalizing of ethical responsibility, that virtues and ethics are closely tied insofar as the presence or absence of both comes to define what man or the soldier is.

DUTY, HONOR, COUNTRY

Virtues that are central to the evolution of military professionalism are those expounded in the West Point motto "Duty, Honor, Country." Duty may be defined as "an action or act that is due by reason of legal or moral obligation; that which one ought to do or is bound to do as an obligation." (Oxford English Dictionary.) Duty consists essentially in living up to one's professional obligations but within ethically acceptable limits. A person is not doing his duty if he becomes like the SS officer who in the execution of his obligations to his orders, his men, and indeed even to the state undertakes unethical acts. As one author has noted, "to educate the military professional is precisely to increase the extent to which the morally appropriate options in difficult contexts are identified and understood."[4] Duty does not consist merely in carrying out the orders of one's superiors or the state, or for that matter even of the profession. Duty consists in fulfilling the obligations of one's profession against the background of a genuine moral sensitivity—against the background of realizing that the obligations of the profession do not constitute the total obligations of the moral man. Thus, in some circumstances an obligation to disobey may arise. It is the realization that ethics consists of recognizing and making difficult choices that forms the background variable against which the virtue of duty must be taught and exercised. Duty is never total obedience; it is only the obligation to obey those orders that are not ethically wrong.

Historically, a good soldier has been defined as one who is loyal to his country, one who keeps faith with his fellow citizens in terms of the oath he has taken to uphold the lawful (and ethical) dictates of the state and, in the U.S. context, of the Constitution. This implies that an effort on the part of any group, even the government itself, to use violence or force in an illegal, unconstitutional, or immoral manner is subject to challenges by the highest authorities of the military profession itself. As General George C. Marshall pointed out, "an officer's ultimate commanding loyalty at all times is to his country, and not to his service or to his superiors." The soldier must realize that, although the military is itself a social institution, it serves the country and the legally constituted civil order first. The soldier's oath is not an excuse for blind patriotism, nor is it an excuse for executing the orders of all civilian authority, no matter how silly or stupid or criminal those orders may be simply because they are issued by a civilian authority. The fundamental concept of loyalty to the country rests in understanding that the military serves the critical function of defending the society against aggression as well as understanding that it is loyal to the larger concept of the nation and its society. The soldier must be aware that there are times when appropriate legal civilian authority can (and probably will) order actions that are devastating to the country. In such instances, the soldier will find himself confronted with a serious ethical dilemma that will require him to choose among very serious obligations. Faithfulness to one's fellow citizens and to one's country implies the ability and willingness to look beyond the short-run policies of particular civilian administrations or regimes and to understand that one's loyalty as a soldier is to the larger values of the nation. Thus, loyalty to one's country can never be allowed to be perverted into blind and total loyalty to superior authority and policies that are immoral or detrimental to the nation itself.

Honor, among its many meanings, has to do with moral sensitivity. Honor is the ability to recognize moral dilemmas and to have the integrity and strength of character to act upon one's perception.[5] It is an integrating trait of the soldier's character, and it prevents the application of technical military skills from becoming an exercise in moral horror. Honor, like integrity, as a perspective of moral sensitivity gives meaning to other character traits of the military professional. The soldier must be aware that he sometimes has an awesome task that involves grave ethical questions. He must also be aware that his own integrity and sense of ethical balance, his honor, is all that stands between him and immorality in the pursuit of this profession. Moreover, as a member of a profession his acts have an influence that reaches beyond himself and affects his fellow soldiers. Honor ultimately rests in moral sensitivity, being aware of the multiplicity of ethical dimemsions to one's actions and being able to act upon them.

In one sense, the motto "Duty, Honor, Country" is reversed. In its present order, it implies that the soldier's first loyalty is to himself as a military professional when, in fact, it is to himself as an ethical being. Indeed, the notion of duty is understandable only in an ethical context. Even duty to his country is not a duty when the soldier is asked to carry out immoral actions. It might be suggested that the motto be turned around to read "Honor, Country, Duty," thereby offering its true meaning within the context of instilling military ethics in the soldier. Honor, as the first trait, implies moral sensitivity which gives meaning to other character virtues; country, as the second trait, implies that the goals of the profession, protection of the Constitution and the nation, transcend duty understood as the rigid adherence to orders; and duty is understood as the obligation to carry out orders only when the morality of the request is understood and judged not to be immoral. This reversal would put the responsibility for ethical judgment squarely on the individual officer as a moral actor. Furthermore, it would emphasize that the commands of duty and civilian superiors are meaningful and valid only within an ethical context. One could, therefore, at least mitigate a tendency toward blind obedience resulting from misunderstood patriotism and avoid, if only in ethical theory, what happened in Nazi Germany when the military virtues of the German soldier were perverted and placed at the service of patently unethical civil and military superiors. But whatever their order of importance, duty, honor, and country lie at the center both of the profession of arms and of the soldier's character.

LOYALTY, HONESTY, SACRIFICE

Every military professional must have loyalty, honesty, and sacrifice. Loyalty can be defined as "the faithful adherence to one's promise, oath, or word of honor." The term "loyalty" derives from the medieval concept of fealty, freely carrying out obligations undertaken by oath. The soldier's loyalty is extracted essentially from the oath he takes upon entering the profession to preserve, protect, and defend the Constitution. Loyalty ought never to be confused with *obsequium*, a perverted loyalty to persons which can obscure one's higher loyalties. At least from the perspective of the profession of arms and the soldiers within it, loyalty to superiors ought never to be so interpreted that it interferes with the larger legitimate loyalty to the Constitution, the civil order, and the profession itself. Loyalty to superiors ought never to extend to covering up incompetence or to preserving a position when to do so has negative practical consequences or when it erodes the soldier's larger loyalties. The point is that loyalty as a virtue rests in adherence to one's promises or oaths. In the context of military ethics, this means that loyalty is extended to faithfully and ethically carrying out those obligations that one has sworn to uphold as a member of the profession of arms. In short, loyalty requires that the code of military ethics be observed as the basis for the soldier's professional actions.

The confusion of loyalty to one's oath and to military ethics with loyalty to individuals is often evident in the military academies. In a study of military loyalty by Lieutenant Colonel John Moellering, this problem is evident in the following description:

> West Point cadets cheered their Commandant, Major General Koster, when he announced his resignation from the Academy citing the charges against him as the commander of the division involved at My Lai. Many doubtless cheered in

affirmation of their loyalty to the Point at a time when it seemed under attack. But those who read or heard of the event could legitimately raise serious questions about the moral discrimination of young men chosen for military leadership.[6]

Thus, in understanding loyalty as a military virtue, it is important to recognize that loyalty is never a substitute for ethical judgment, nor can it be used to abandon ethical judgment or as a defense for refusing to make ethical judgments. Loyalty properly understood implies that the soldier's fealty is to his oath, the Constitution, and his sense of professional ethics, and only secondarily to his superiors. Moreover, loyalty to anyone or anything can never be used as an excuse not to discriminate ethically. Loyalty in following orders is not an excuse for the soldier's failure to observe his larger obligations which entail the observance of the code of ethics.

Honesty is crucial for the military profession, for falsehood creates the possibility of drastic miscalculation wherein the lives of men are spent in vain. Honesty has been defined as "uprightness of disposition and conduct; the quality opposed to lying, cheating, and stealing. Honesty is honor gained by action or conduct." A soldier who is not honest is a greater liability than asset, and, in some circumstances, he is a danger to himself and his comrades.

It can be argued that honesty is more important to today's soldier than to warriors in the past because of the vast destructive power of modern weaponry. The number of lives that could be lost and the consequences to oneself and one's comrades as a result of failing to tell the truth are enormous. Samuel Hayes says in this regard that, "lives, careers, battles, and the fate of nations have hung on the ability of military leaders to state all the true facts to the best of their knowledge regardless of what effect these facts may have on themselves or others."[7] From this perspective, honesty is crucially important to the development of a good soldier, for if a soldier cannot be relied upon to tell the truth and to be honest in his dealings with his fellow soldiers, superiors, and subordinates the consequences are practically and ethically devastating. A dishonest soldier, especially a dishonest officer, has no worth.

With regard to sacrifice, it is the very basis of professionalism. The military is sworn to serve the state and the society. This inevitably means that at some point the members of the profession will have to pursue the interests of their client instead of their own. As noted earlier, this obligation is clearly reflected in the clause of unlimited liability. As harsh (or idealistic) as it sounds, the truth is that the soldier may be legitimately asked and required to make the ultimate sacrifice of his own life in observance of his professional obligations. Sacrifice is a noble virtue when it is done for values that are worthwhile. There is no virtue, except perhaps the virtue associated with theater, in pointless sacrifice. If it is understood that the nature of the profession is to render service and to act ethically in the rendering of that service, then the sacrifice of the soldier's life either for trivialities or in pursuit of immoral policies is not only pointless but may in fact be wrong in itself. Consider, for example, the Nazi regime wherein soldiers sacrificed to perpetuate a gang of thugs who themselves had no sense of morality. To die or suffer for such values is pointless.[8]

Those who enter the military profession are expected to demonstrate the trait of sacrifice, although the giving of one's life is not always involved. At the minimum, they must be prepared to forego their own self-interest in the service of the larger good of the profession and the society they serve. Moreover, they must be aware that if circumstances warrant the contract of unlimited liability may come due and they will be expected to pay in full. This sense of virtue is not easily developed or sustained, nor indeed is it developed without an unambiguous understanding as to what it implies in terms of the price the soldier may have to pay. Members of the profession who strive to develop the virtue of sacrifice and who succeed in approaching it deserve to be regarded as among the most noble of their fellow soldiers.

PATRIOTISM

Patriotism is defined in the OED as ''the quality of disinterestedly or self-sacrificingly exerting one's self to promote the well being of one's country; one who maintains and defends his country's freedom or rights.'' Life in the military extracts far more in the way of obligations than it returns in the way of benefits. Men in the profession of arms are expected to be true patriots, giving service to their country out of affection for it at costs that are largely absent from occupations in civil society. Patriotism implies a love of one's country that is founded in clear notions of ethics. Patriotism does not imply a blind loyalty to the state regardless of its conduct; rather, it implies that a man is prepared to serve the state in its pursuit of those tasks that have strong ethical foundations. In this sense, patriotism can never be, as Samuel Johnson suggested, the last refuge of scoundrels, especially for the soldier. Those who do the bidding of an immoral political order are not patriots; they are accomplices. Patriotism is no refuge at all. It is the first place where men of honor, men of vocation, are found willing to sacrifice in defense of a country whose policies and objectives are morally tolerable.

OBEDIENCE

Earlier, a distinction was made between obligation and obedience. Obligation, it will be recalled, was defined as observing those precepts that one understands and willingly agrees to observe for the reason that they ought to be observed, given the context in which they arise. Obedience, on the other hand, means complying with the will of another, even if one does not totally understand all the reasons why certain instructions have been given. The concepts of obligation and obedience are often confused—so much so that some contend that encouraging a sense of obligation based on an understanding of ethics produces a tendency for the soldier to be disobedient. The argument misses the point.

Not every disagreement or different point of view or different perspective surrounding a question or problem raises an ethical question or constitutes a moral dilemma. Nor, indeed, are all obligations moral obligations. For example, the obligation that a good soldier ought to ensure that the tanks are painted every month is not a moral obligation in the normal sense of the term (although perhaps it could become so in the extreme). It is simply an obligation. Obedience involves executing legitimate orders, even when one disagrees or does not understand the reason for them, as long as they do not raise concerns of ethics in the minds of those who are expected to obey. To be obedient to the will of one's superior, even in the face of disagreement, does not necessarily involve an ethical conflict since many of the orders to be executed do not involve ethical questions and may have no ethical content at all. Expressed in terms of a sports analogy, a quarterback may call for a play with which his halfback disagrees but which requires that the halfback execute the play. In these conditions, it is fully expected that the halfback will obey and attempt to carry out his instructions to the best of his ability. Only when the quarterback orders the halfback to throw the game does a simple question of obedience become transformed into a major ethical dilemma requiring the halfback to make a clear distinction between obedience and obligation.

A line must be drawn between orders involving ethical obligations and those requiring obedience. A good soldier is an obedient soldier who enthusiastically carries out the orders of his superiors even, at times, if he does not like them, agree with them, or even understand them fully. However, the soldier's virtue of obedience should never be taken to mean that it condones or allows him to abandon his ethical obligations in any sense whatsoever. When the orders of one's superior raise an ethical question in the mind of the soldier, he is obliged to solve the ethical dilemma *before* carrying out the orders. The solution may indeed require that he refuse to carry them out, but that is the very nature of ethical judgment. Obedience to orders is never a substitute for ethical judgment. At the same time, the requirement of obedience does not always involve or bring to the fore questions of ethical judgment or ethical obligations.

DISSENT

Not every directive or order involves moral questions, but when they do, the soldier has certain basic obligations. Among these is what a former commandant of the Naval Academy called the "will to dissent."[9] A good officer is courageous enough to disagree with his superiors when he feels the issue involved is important; he dissents openly from the judgment of his superior. Dissent is the opposite of the "CYA (Cover Your Ass) Syndrome" in which officers wish to "go on record" as upholding a certain "position" so that if things go wrong they can point to their boxes of "memoranda for record" in order to escape responsibility for the consequences of failure. The CYA Syndrome is a perversion of the virtue of dissent.

Dissent implies a soldier's willingness to explore with his superiors the rationale behind their directives so as to better understand the basis for them or to point up the difficulties he perceives to be associated with them. The object of dissent is to bring out a point of view which the soldier believes may not be evident to his superior or, perhaps, is not given sufficient weight in the decision. There is never any question that the point of dissent is to try to help one's superiors make the best possible decision under the circumstances. An officer should be encouraged to dissent and to develop the "will to dissent" as part of his character.

The Character of the Soldier   163

One historical figure noted for his dissent from his superiors was Admiral Lord Nelson, and he encouraged his young officers to do likewise. Prior to battle, he would assemble his officers in the wardroom and ask those present what they thought he ought to do. He always required his junior officers to answer first, so that they would not be cowed or have their views influenced by those of their seniors. Nelson felt strongly that a good officer would be willing to offer a contrary point of view. Nelson's habit had a profound influence on the famous Japanese Admiral Isoroko Yamamoto who required the same practice of his officers during World War II. Dissent should never be allowed to degenerate into mere carping or bureaucratic self-defense, tactics whose objective is to allow the soldier to escape responsibility, not enhance it. The object of dissent requires that one accept responsibility for one's contrary point of view. Hence, a good soldier is obedient and yet is willing to dissent when he feels it is appropriate to do so. "Yes men" never make good soldiers or officers. As a military professional, the soldier must also learn to assess when dissent is appropriate.

The dangers normally associated with dissent in the military environment are overemphasized. It is highly unlikely that dissent within the military would ever degenerate into wide-scale emotional debate on any issue and thereby bring the profession to the point of paralysis. The greater danger in any organization, especially one organized as bureaucratically, hierarchically, and authoritarianly as the military, is exactly the opposite, namely, that a lack of dissent will permit the perpetuation of failing or unethical policies. Examples include body counts, pacification projections, indicators of combat performance, falsified intelligence and readiness reports, and ordnance expenditure doctrines, all of which were common failing policies during the Vietnam War. The perpetuation of failing policies over a long period is possible only when soldiers and officers do not possess the will to dissent. An officer who is unwilling to dissent is an officer who is unsure of his own competence and, perhaps, even his ethical moorings. Dissent as a mechanism for helping one's superiors see things as they really are and to reach the most effective decisions possible should always be encouraged by military training.

ABILITY TO LISTEN

A good officer must be willing not only to dissent, but also to be a good listener. He must be willing to explore all areas and facets of a question and to gather all relevant data by listening to others before finalizing his own views. Moreover, he must be willing to alter his positions if any new data come to light which shed doubt on his initial views. In a society that emphasizes expertise and specialization, trends evident within the military itself, the tendency is to defer excessively to expertise. This deference should never become a substitute for independent judgment or for listening to dissenting views. Nor should the soldier ever allow expertise to blunt his questioning of the basis of any decision. The soldier must be able to grasp the subtleties of the arguments of others and not to simply rebut them (although that may be required at times) but rather, thoroughly understand them and the role they play in the decision. The imperative to learn is no less important for the soldier than for anyone else.

INTELLECTUAL CURIOSITY

Soldiers must be thinkers infected with an intellectual curiosity. Action is, of course, required of all soldiers, but action without thought and plan is likely to produce a disaster. Ethical decision-making also requires thought, a willingness to explore all aspects of the circumstances in which one must make decisions. A good soldier must be able to improvise as circumstances warrant, and that requires that he think and reason about those circumstances. It is simply not true that discipline is the antithesis of thought. True discipline is the steady application of a course of action carefully reasoned out. Thus, soldiers, especially officers, must develop intellectual curiosity and a willingness to explore areas that are unfamiliar to them in order to obtain a better grasp of information relevant to their role as military professionals. The soldier must have a genuine desire to know, and to learn about new things. A good soldier must be willing to exercise his mind, to think the unthinkable if necessary, and to trust his judgment.

MORAL REASONING

To suggest that the soldier must be a thinker and a good listener, and possess the willingness to dissent is to imply that he must have the virtue of moral reasoning. A soldier cannot act ethically without moral reasoning. If the essence of ethical choice is the ability to choose one obligation over another when one cannot do both and to know the reasons that underlie one's choice, then moral reasoning rests at the center of ethical action for any moral agent, including the soldier. One of the major shortcomings of military education is a demonstrable tendency to avoid serious instruction in moral reasoning. It is assumed instead that a soldier already brings with him an ability to reason morally. In most instances, the techniques of moral reasoning are developed almost like an art, and the notion that most soldiers bring this art with them when they enter military service is generally not justified. Moral reasoning occupies an important place in the character of the soldier, and it is a primary task of the military profession to inculcate and develop this virtue as much as possible.

RESPONSIBILITY

If one were forced to choose the one virtue that the soldier must demonstrate above all other military virtues, many would choose a sense of responsibility. Responsibility involves first the understanding that an individual be his own agent, responsible for his own actions. To accept responsibility means being willing to accept the burden of ethical decisions and to accept the consequences of those decisions be they good or bad. Fundamentally, responsibility implies that the soldier recognize that to be human is to be obligated and that to make hard choices among less than ideal options is part of trying to act ethically. It is in the recognition of the special tasks as they affect the soldier that he truly demonstrates responsibility. An officer who tries to avoid responsibility of any kind, especially ethical responsibility, is likely to become a managerial careerist. Thus, he is not likely to contribute to the moral development of the profession, to say nothing of enhancing his own moral development. A soldier who will not bear responsibility is not a military professional.

HUMANISM

As a humanist, a soldier must understand that what he does, while valuable in itself, cannot be isolated from its impact on other human beings. He must understand that, while the acts of a good soldier might be approvable in a narrow sense, there are times when to be a good soldier is to be the antithesis of a good human being in the larger sense. He must possess a scope of obligation and ethical awareness that transcends narrow professionalism and includes the realization that the application of his skills can have terrifying effects on his fellow man. From this perspective, he must also bear a sense of social responsibility. He must understand that he is not an island, that what he does affects not only himself but also other men within the profession and even the society as a whole. Top-ranking military men must be especially aware of the responsibilities they have to the larger society, for upon their recommendation it is possible to sentence whole societies to death. An awareness that the military performs its tasks within a social environment and interacts with that environment constitutes a virtue for the soldier.

COMPASSION

A soldier, aware that he deals in life and death and that his decisions may unleash terrifying violence upon his fellow man, must develop compassion for human suffering. He must be capable of genuine compassion for the suffering and distress of others, and he must have a desire to relieve that suffering if he can.

A doctrine of "minimal application of force" is also ethically applicable to the soldier. He must apply violence in the pursuit of his profession, but he ought never to revel in the effects of violence upon other human beings. This is not to say that he cannot be a proud professional in the application of his expertise, only that he must understand that destruction of human life is a less than ideal means to sometimes good ends. The soldier must feel some hesitation at willing the destruction of human life; he must not do it too readily or enjoy it too much. The soldier must have a willingness to stay the hand of the sword when it is possible to stay it.

REALISM

A soldier must be able to make an honest and impartial assessment of the application of rules for himself, his men, and even the enemy. A good soldier is realistic; he must understand that in an imperfect world only imperfect solutions are sometimes available for perennial problems. He must develop a sense of realism that helps him realize his own limits in the application of his military skills and to understand that man is limited in his ability to grasp even the consequences of his own action completely, to say nothing of the consequences of the acts of large-scale organizations dedicated to the application of systematic violence. Although a realist, the soldier must resist becoming a narrow professional, resist the pull of exaggerated self-interest, and resist the tendency to "get ahead" at any cost. Through realism, soldiers come to understand that they cannot do it all, that they cannot be perfect, and that the real world is almost never fair and seldom lends itself to easy solutions to most problems. Yet, a sense of realism should not lead to a feeling of despair; it should only lead to a careful balancing of means and ends and to a careful selection of ends that are possible to attain. If realism is allowed to degenerate into despair, it paradoxically becomes an unrealistic way of seeing the world. Even in an imperfect world it is possible to try to achieve conditions and objectives that are in themselves worthwhile. Nonetheless, an officer who is unrealistic in his choice of goals or in the way he selects his means will not be of much help to his superiors or to the men he leads. Thus, humanity, social responsibility, compassion, fairness, and concern must be balanced by a concern for what is possible in an imperfect world.

VOCATION

For the profession to take on an aura of a vocation, a sense of special obligation is required. The soldier must realize that what he does is categorically different from what other men in other occupations do, and that the obligations he undertakes are also different and special in their scope and level of responsibility. A soldier who enters the military only for career goals will rapidly discover in the carnage of the battlefield how disjointed his priorities are. No sane or well-balanced human being can be expected to endure the terrors of battle and the responsibility of sending men to their deaths, to say nothing of inflicting death upon other human beings, simply in pursuit of his own power, prestige, income, and career status. These factors which supposedly motivate men in the civilian community are insufficient to motivate the soldier on the battlefield. The only thing that will prevent soldiers from degenerating into hired killers or armed thugs is an awareness of their special obligation, and this special obligation rests at the center of military professionalism.

What is required among military men is a sense of brotherhood, of understanding that one's fellow soldiers within the profession share the same obligations, the same risks, the same costs, and, one would hope, the same virtues. Members of the military are bound together by their willingness to assume and pay the price of belonging as members of a special social group.

DEDICATION

Soldiers who share a special sense of obligation as in a brotherhood must also be dedicated. They must demonstrate the virtue of dedication, a term that comes from the Latin *dedicare*, "to proclaim or to devote or to consecrate." When a soldier is said to be dedicated, he is said to have a sense of being set apart. The soldier must be willing to devote much of his life to a purpose beyond his own self-interest, his sense of brotherhood and obligation transcend this interest. One becomes a dedicated professional when one understands, recognizes, and assumes the special tasks, burdens, responsibilities, and obligations of the profession. One is *not* a dedicated professional when one merely pursues one's career self-interest defined in terms of promotion, rank, and good assignments. The Pentagon is full of soldiers who have had good careers. Its saving grace is that it is also full of good soldiers who have not had successful careers but who still remain good soldiers. Dedication means a devotion "to a sacred purpose with solemn rites," and accordingly it implies a devotion to something beyond one's self, to the community and the profession of which he is a member.

INTROSPECTION

If we expect soldiers to feel a sense of dedication and to buttress that with a sense of brotherhood we had best be prepared to require of them the virtue of introspection as well. It is no accident that monastic orders require of their members that a certain period of every day be reserved for introspection. Introspection should also be an essential part of the mental training of the military man. The German Army, when it adopted the general staff system in the mid-nineteenth century, institutionalized the concept of introspection in the practice of the tactical walk. In effect, it sent an officer away to another assignment or on a leave of absence and required him to think about a solution to a complex problem or to pursue a course of study, often philosophy, purposely unrelated to his military specialty.

Introspection is so crucial because the military man can easily be lured by the prestige associated with rank and promotion. Almost by accident, he can become so wrapped up in career that he can lose his ability to think independently. The ethics and virtue of the military profession as a special calling are thereby threatened. In reacting almost exclusively to external conditions, a soldier increasingly begins to respond to these conditions as if they were singularities. A soldier who cannot look inward will quickly find himself pursuing goals that are almost always materialistically based.

Through introspection, the soldier can maintain his ethical balance and his sense of special obligation which make him a true soldier. If the soldier starts to lose this sense, introspection can lead him back to the right path. The virtue of introspection constitutes an application of an old monastic principle to the modern military profession; it is perhaps even more valuable, more needed, and more appropriate in an age of increasing complexity.

IMAGINATION

Imagination does not readily come to mind as an important quality of the soldier, for he is stereotyped as a fellow immersed in a large organization run by rigid rules and orders that allow no deviation. Yet, military men themselves always rank imagination very highly. The soldier requires imagination if he is to be effective on the battlefield. As General Graf Moltke was fond of pointing out to his staff, there was never a military plan that could survive twelve hours' contact with the enemy. Thus, a good officer uses his intellect to generate new options not covered by regulations, plans, and standing orders. This lesson is applicable to the soldier as well. The conditions of virtue are essentially conditions of the mind, and the quality of imagination is a vital one in any officer. Nothing could be more false than to believe that an ideal soldier is a man addicted to regulations and following only the proven path. The best officer is likely to be the one who "makes his own tracks" as one military commander, undoubtedly an armor officer, has chosen to express it. Indeed, without imagination many of the other qualities of a good soldier might be easily squandered.

CONFIDENCE

An officer's virtues and technical skills will count for little if he does not have the confidence in himself to execute his responsibilities boldly.

Without confidence, he will not be given to risk or daring or any of the other qualities so vital to behavior in battle. More than that, lack of confidence is contagious. An officer who does not radiate confidence will quickly find that his men lack confidence in him.

COURAGE

Physical courage in the face of fire and hardship is, of course, required of the soldier if he is to be a good leader. Most officers in a normal career, however, are unlikely to find themselves engaged in combat for more than a relatively short period of that career. Their daily activities will require another kind of courage—ethical and moral courage. Ethical courage requires a willingness to deal with difficult situations without fear, to accept the risks and responsibilities, and, if need be, to be willing to bear the cost of a course of action that one believes is right. Without physical courage, a soldier cannot be an effective combat leader; without moral courage, he cannot be an effective officer during times of peace or war.

BATTLEFIELD VIRTUES

Military men writing about military virtues tend to divide considerations of military virtue into those that are most important on the field of battle and those most appropriate in the normal noncombat environment. More often than not, the division is unconscious, although it remains a distinction that emerges constantly throughout military writing. In biographies and memoirs, military men most often cite the following virtues as most desirable in the men they command on the battlefield: aggressiveness, the tendency to attack, daring, risk-taking, initiative, perseverance, and flexibility. Molding these qualities together is the virtue of perspective—an ability to assess risks as they relate to initiative and daring, not to squander one's men meaninglessly in the pursuit of personal bravery or glory but to be aware of the mission requirements relative to the real human costs.

One important battlefield virtue is discipline. In common usage, to be disciplined is to be dedicated to a course of action that is beneficial to attaining the goals of one's life or profession. The discipline of an officer is important to maintaining the discipline of his men. And disciplined troops are vitally important to good leadership and military effectiveness. Discipline is also required if men in battle are to act ethically, to minimize the damage and killing. Only when the mind is disciplined can it think and make choices essential to the proper conduct of military ethics.

BEARING

The bearing of a soldier establishes a reference point between himself and the ethical code of his profession expressed in terms of what he is supposed to do as a good soldier. Equally important, bearing establishes a reference point between a soldier and his fellow soldiers and, perhaps even more crucially, the men he leads. He becomes their compass by providing an example for them to emulate or, at least, follow. A combat leader sets the example of ethical behavior for his men. A leader must convince his men that he is deadly serious about the ethical limitations of combat, and by his example and other means he must ensure that they observe them.

DECISIVENESS OF CHARACTER

In simple terms, a soldier must be prepared to decide. An officer can be forgiven making the wrong decision, but he will not be forgiven the unwillingness to decide. Military life requires that soldiers make decisions and that they accept responsibility for them. Decisions are made more frequently in the military than in other professions. And, of course, military decisions are frequently more important than those made elsewhere. Many of these decisions deal with material things, but all too often they involve the lives and deaths of men and have to be made with incomplete information and very quickly. In addition, the soldier's decisions often put his own life at risk, a condition not found in other professions. Thus, a soldier who will not or cannot decide is useless. A good officer must be able to decide and to make his decisions felt with a sense of confidence and aggressiveness if he is to be an effective leader of men.

DIGNITY

A soldier must have dignity of character. The term "dignity" is derived from the Latin *dignus*, meaning worth. A soldier, then, must have worth, and he must demonstrate his value to his men and to his peers. Like the Medieval knight there is a sense in which the soldier is different. He ought to be the personification of a set of virtues, ethics, and values which constitute a *raison d'etre*. This way of the soldier is demonstrably different and separate from the larger society. The officer must demonstrate his self-worth, a sense that what he is doing is truly important, and he must be able to transmit this sense of worth to his men. Even as an individual standing alone on a desert plain he must exude dignity, a sense of being above the normal daily marginal concerns of human existence, and in his bearing and dignity he must reflect the worth of the profession of which he is a member and which he serves selflessly.

TECHNICAL COMPETENCE

Demonstrating an acceptable level of technical competence is a problem that occurs more frequently in modern armies as they become more complex and specialized. Officers now tend to be narrowly competent in one area rather than broadly competent in many. The artilleryman, for example, who has no idea of how to disassemble a rifle and the rifleman who has no concept of the applications of armor or artillery are classic examples. Fortunately, the military is making a serious effort to create within the profession a general sense of broad professional competence to counter the narrowness attendant to specialization. Thus, most officers who have attained the rank of general officer have had career patterns that have forced them to circulate through a series of different assignments in an attempt to generalize their experience and to wean them away from a narrow technical specialization. Technical competence implies not only competence in the skills one is most likely to use in any given assignment, but also an ability to become functionally competent in a number of areas. The future still belongs to the generalist, not the technician.

POWER

Military effectiveness is only tangentially related to technical competence in the sense of troops possessing certain military skills. Rather, combat effectiveness seems to be fundamentally related to cohesion. Units that have no cohesion are not likely to fight well regardless of the quality of their weaponry or the degree of their training. Moreover, building military cohesion is a function of the interpersonal relationships that develop between the leaders and the led. Military effectiveness, therefore, depends heavily upon the extent to which an officer is capable of creating, developing, and sustaining those interpersonal skills that allow him to build strong ties with his men. Without strong bonds among men in combat units, these units will not withstand the stress of the modern battlefield.

The officer who holds a position of command also holds a position of power, and he must never forget that effective leadership is closely bound up with the application of power in a social sense. As an officer and a soldier, he is responsible for developing his skills and fostering the bonds of cohesion that make military units operate well in battle.

Conclusions

The list of virtues presented here, as long as it is, reflects the connection between the special responsibilities of the military and the character required of the men to bear them. No one can realistically expect soldiers to demonstrate all of these qualities of character all of the time. The notion of perfect virtue is not applicable to the imperfect empirical world. Yet, the list of military virtues discussed here constitutes an adequate list of ideal conditions of character worth striving for. They can serve as guidelines for the soldier's actions as a military professional.

Virtues in themselves are not the equivalent of ethics or ethical action. All too often the military academies and the higher level staff colleges have tended to believe that the inculcation of personal virtue will itself provide the soldier with adequate precepts for ethical action. Most certainly, one cannot be expected to act ethically in the absence of at least some of these virtues, for virtues represent predispositions to action. But they are not action. They are not specific in their applications of what one ought to do, especially so for the soldier and the varied circumstances he is likely to confront. Only a code of military ethics can provide specific precepts as requirements of action. But a code of ethics would be meaningless for a soldiery that did not possess some military virtues because none of the predispositions to act ethically would be present. Thus, a code of military ethics as well as education and training in the development of military virtues go hand in hand in producing an ethical soldier.

Finally, military men who lack virtue or ethics can never be adequate servants of society or the profession they are sworn to serve. As S.L.A. Marshal has pointed out, "To the extent that military men lose their faith in virtue and become amenable to ill-considered reforms simply to appease the public, they relinquish the power to protect their society." It must be affirmed once again that the way of the soldier is different from the way of the civilian, and the military's requirements for virtue and ethical action are also different. Those called to the profession of arms are required to bear heavier and different burdens from those of their fellow citizens outside the military. But it is in assuming these burdens that citizens can truly become ennobled. If the soldier does not do his job well, no one in the larger society can hope to benefit from the advantages of the society. It is a curious irony that the pursuit of self-interest can only be made possible when a group of special men is willing to forego that pursuit and defend the society so that others may engage in it. The very survival of society and its quality of life depend on the presence of an adequate military force. And an adequate military depends on a core of professionals who themselves are virtuous and who are willing to observe a code of ethics that sets them apart from their fellow citizens while placing themselves in their service.

Notes

1. Maxwell D. Taylor, "A Professional Ethic," *Army* (May 1978): 20.
2. This is a dictum of Kant's as quoted in William K. Frankena, *Ethics* (Englewood Cliffs, N.J.: Prentice-Hall, Inc., 1973), p. 65.
3. Quoted in Henry J. Meade, "Commitment to Integrity," *Air University Review* (March-April 1977): 88.
4. Arthur J. Dyck, "Ethical Bases of the Military Profession," *Parameters* (March 1980): 44.
5. Ibid.
6. John H. Moellering, "The Army Turns Inward," *Military Review* (July 1973): 68.
7. Samuel H. Hayes and William N. Thomas, *Taking Command* (Harrisburg, Penn.: Stackpole Books, 1967), p. 51.
8. Kenneth H. Wenker, "Professional Military Ethics: An Attempt at Definition," *Air Force Journal of Professional Military Ethics* 1 (April 1980): 25.
9. Ibid.

7 LOYALTY, OBEDIENCE, AND DISSENT

For the military professional, the question of military obedience extends also to the limits of his obligations to follow the orders of his military superiors. Understandably, the military is very sensitive to the question of its obedience to civilian authority. In most instances, the question is never raised formally in the curricula of the military academies or staff colleges. Indeed, one gets the impression that the military assumes there are no practical limits to its obligations to obey legitimate civilian authority. Thus, one way of dealing with this vital question of politics and ethics in a democratic society is simply to ignore it, implicitly affirming that the military's obligation to follow the dictates of its civilian overseers is absolute. Yet, there *must* indeed be limits, even in a democracy. The idea of legitimate limits has been accepted as a principle throughout military history in the American military and in most armies of the Western world. As a necessary corollary, it has long been held that there are discernible limits to the obligations subordinates legitimately owe to their military superiors. It seems appropriate, therefore, to address the question of loyalty, obedience, resistance, and dissent within the military profession as legitimate avenues of ethical action for members of the profession.

There is not the slightest hint of evidence in all of American military history that the military has ever considered violating its constitutional limitations. In addition, a number of recent studies, including the already noted Squadron Officer Study and the Army War College Study, reveal absolutely no support for the view that the miliary should "instruct" its civilian superiors beyond the limits specified by the Constitution. There is no danger that the dogs of war wish to or will slip their leash.

Nonetheless, the military's oversensitivity to the issue is understandable from the political standpoint in that the question of limiting its obligations to obey inevitably raises in the minds of its political masters the spectre of a "state within a state." But in failing to deal with the question adequately, a major ethical issue has been left unaddressed, and so the military has not specified any ethical precepts that could serve as guidelines for actions in the event circumstances forced the question into the open.

A False Fear

Despite sporadic fears to the contrary, there is no evidence that the military would support the creation of a state within a state. In fact, the evidence strongly suggests the existence of strong opposition within the military itself to the slipping of its constitutional bonds. In numerous conversations with officers who served through the last days of Watergate when it was rumored that the president may have become unstable, many officers noted that for the first time in their careers they thought seriously about what they would do if their commander-in-chief ordered them to take actions they knew to be illegal or unconstitutional. I have been unable to discover a single officer who was prepared to follow such orders. Indeed, in a splendid display of judgment and patriotism, the military itself initiated the question of its limits to obey their commander-in-chief. The Joint Chiefs of Staff raised the question with the then secretary of defense, James Schlesinger. They formally requested that the secretary clarify the line of the chain of command as specified under the Defense Act of 1958. That act requires all orders of the president dealing with the deployment of troops to be passed through the normal civilian chain of command beginning in the office of the civilian secretary of defense. In short, they were reaffirming the legal groundwork for not executing orders given directly to military commanders by the president which bypassed the secretary of defense. The Joint Chiefs themselves sought to clarify their moral position, as did hundreds of their brother officers throughout the profession, as to what they would do if they were ordered to undertake actions they thought to be illegal. While history never reveals its alternatives, the consensus that one gathers through conversations with serving officers is that most, including their highest ranking superiors, would not have executed any such orders. Given this experience, perhaps unique in our history, it seems safe to suggest that the idea of exceeding its constitutional limits is as loathsome to the military as it is to their civilian superiors. The American military is fundamentally committed to its limited constitutional role, and the fear of excessive military influence or threat is a much inflated fear.

There is no fundamental contradiction between the existence of a military organized essentially along authoritarian lines, whose values are separate from society, and the propensity for the military to constitute a threat to the democratic civil order. One can point to England as a democratic civil order of long standing whose military has never been a threat to its internal regimes. The same tradition exists in Canada and has existed in the United States since its birth. The tradition is clear, if somewhat less strong, in France. Even the Germany Army has not represented a threat to its civilian leadership historically; indeed, its failure has been its tendency to acquiesce too willingly in the policies of any civilian government. Israel is a classic example of a liberal democratic state whose authoritarian military presents no threat to it. An examination of the military organizations within the Western world over the last two hundred years shows no clear examples of a military prepared to exceed its defined role and to subvert or remove constitutional civilian authority. In the United States where the military has never been a threat to the democratic order, what exists is a fear of such a threat, a fear without any real foundation.

The real problem of military structures in the West, especially in the United States, has been the demonstrated propensity not to question civilian authorities, even when their orders run against the best judgment of military professionals. The American military has a long history of compliance, even when it considered the policies it was expected to execute to be wrong, failing, or indeed, unethical. The evidence, for example, is fairly strong that much of the military was opposed to most of the strategic and tactical policies adopted in Vietnam.[1] They simply never spoke up!

The creation of a professional military is likely to enhance civilian control by narrowing the sphere of legitimate political activity for the military rather than decreasing that control.[2] The most frequent problem is not that the military has sought to increase its role at the expense of civilian control, to usurp legitimate authority, or to act beyond the range of the Constitution. Rather, it is the opposite tendency to carry out the orders of civilian authority, at times against the best military judgment.

From this vantage point when one discusses dissent, loyalty, and the limits of military obligations, the central problem is that the military represents a threat to the civil order not because it will usurp authority, but because it does not speak out on critical policy decisions. The soldier fails to live up to his oath to serve the country if he does not speak out when he sees his civilian or military superiors executing policies he feels to be wrong.

The Limits of Obligation

The individual is always at the center of ethical responsibility and ethical action. In this view, judgment and choice are central to ethics, and no soldier can ever abandon his obligation to act ethically to any other man without ceasing to be ethical himself or, indeed, without ceasing to be somewhat less human. An individual is never justified in acquiescing to orders he judges to be immoral, no matter whether they are issued by military or civilian superiors. This is not to say that he may not obey orders of which he is not sure, although he will be held responsible for the consequences. But if a soldier is convinced in his judgment that an order he is being told to execute is immoral, he may not abandon the ethical obligation to resist or refuse these orders in an appropriate manner.

As a profession, the military has the same obligation. As a profession dedicated to selfless service to society and the Constitution, the military cannot simply acquiesce without dissent, without protest, and perhaps without public outcry in any order that is immoral. The fact that the order may have been given by legitimate civilian authority makes no difference. It must be remembered that law has no necessary moral content. No one would seriously argue that a soldier ought to follow the order to kill large numbers of people just because the court that ordered it was legally constituted in following the dictates of the state. To do so would not in any imaginable sense be construed as moral action. Thus, it is a well-established principle of ethics that men cannot abandon their ethical judgment to other men and thereby escape ethical responsibility for their actions. So, too, soldiers as individuals and members of a profession cannot escape responsibiliy by acquiescing in immoral orders, even if they are legal ones.

The military must be loyal to its civilian superiors, and soldiers must be loyal to their superiors in the execution of orders that are ethically and legally right. In terms of the military profession, this obligation is far more to the Constitution and the society than to the transitory occupiers of positions of civilian authority in the Department of Defense or even the White House. The military has a special obligation to bring its expertise to bear in defense of the common good of the republic it is sworn to defend. It also has an obligation, as its oath specifies, to preserve, protect, and defend the Constitution. At its extreme, this obligation implies that if and when civilian authorities undertake actions requiring that the military violate the Constitution, the military has an obligation to resist.

Faithfulness to one's fellow citizens in terms of the vow to uphold the Constitution means that the effort on the part of any group, even the government itself, to advocate or use violence in an unconstitutional manner is subject to challenge by the highest authorities within the military profession itself.[3]

Hence, there are limits to the obligations the military has toward its civilian superiors, and they are drawn very clearly on legal and constitutional grounds. They are drawn less clearly on ethical grounds because of the requirement for individual judgment that may run contrary to the dictates of the state as law. But they are no less real or compelling. This same principle applies to subordinates and their relationships to superiors. A soldier has a legitimate moral and legal obligation to observe the orders of his superiors, providing such orders do not violate the ethical sense of the soldier required to carry them out. When they do, the subordinate has an obligation to question and to resist those orders in an appropriate manner. Thus, members of the military must recognize that there are limits to their obligations to their civilian superiors, and there are limits, too, to their obligations to their military superiors. These limits are specified in terms of challenges to the Constitution, challenges to the very existence of society and the common good, and, most commonly, challenges to the ethical sensitivities of the men who must decide whether to obey or resist.

While there is no question that civilian authorities should control the military and that members of the profession have an ethical, legal, and moral obligation to execute the lawful and moral orders of their superiors, it is important to undersand that this obligation is not open-ended; there is no ethical clause of unlimited liability to obey. When the soldier confronts this obligation, he must be prepared not to follow those orders that are unethical, immoral, or illegal. This counsel of dissent and resistance has a long tradition in American history. The doctrine of limited obligation to immoral orders is found in the military's own manuals. FM-27-10, the Army's manual for "The Law Of Land Warfare," contains the following:

> The fact that the law of war has been violated pursuant to an order of a superior authority, *whether military or civil*, does not deprive the act in question of its character of a war crime, nor does it constitute a defense in the trial of an accused individual, unless he did not know or could not reasonably have been expected to know that the act ordered was unlawful.[4]

The individual who complies with immoral or unethical orders may in fact be prosecuted as an accomplice. General George Marshall observed that "an officer's ultimate commanding loyalty at all times is to his country and not to his service or his superiors."

Military regulations make it quite explicit that the soldier must be prepared to question, and even openly disobey or refuse to comply with, orders he regards as immoral. According to Professor Toner:

> Because of its belief in a higher morality in a God, the American nation can hardly do other than to recognize the soldier's conscience as complementary to and not necessarily destructive of military discipline.... Should circumstances require him to choose between the two, honor obliges him to be responsible for his actions and to accept willingly and manfully the consequences of his choice.[5]

There is no escaping moral choice or ethical judgment in deciding how one should act. This is true in all spheres of human endeavor, and it is most clearly true in the choices soldiers may have to make.

The question of the limits of military obligation raises a problem that most soldiers have great difficulty in dealing with in the pragmatic world. Josiah Bunting points out that members of the military profession seem to have great difficulty recognizing ethical dilemmas and resolving them satisfactorily. Part of the reason is that they receive very little, if any, training in the formal ethical precepts of the military profession, and few ever receive any training in the critical virtue of moral reasoning.[6] In addition, there are strong institutional pressures to avoid dealing with

Loyalty, Obedience, and Dissent 181

problems in ethical terms because they work against a soldier's superiors. They also create an enormous problem of conscience for most men. Finally, the soldier's training and experience militate against framing questions in ethical terms. As Bunting states:

> What does the professional officer do when his conscience troubles him, or even when his intellect alone troubles him, or when the two of them together tell him that the institution of which he is a part is making a very serious mistake? Can he stand up within the institution, make his criticism forthrightly, dare to hope that they will be scrutinized dispassionately and acted upon in a way which may vindicate his judgment? Can he do this without serious risk to the successful development of his career? Generally the answer to both questions is no. Even more depressing is the fact that the problem rarely surfaces.... Everything in the professional soldier's training runs counter to his even posing the question.[7]

Thus, to raise the question of the limits of the soldier's obligations is to levy an enormous burden on men who are already overburdened with special obligations. It is necessary nonetheless, for only moral integrity can truly guarantee the democratic rights of a free society. Only a truly ethical military profession can ever guarantee that the military itself will not abandon its high ethical responsibilities of service and degenerate into a group of armed thugs prepared to use social violence for their own narrow ends. In this sense the doctrine of limiting one's obligations as a soldier applies in the same sense that one has limits to all other obligations. And it is this ethical sense of limit which is really at the foundation of the guarantees that the military will remain loyal in its proper social role. This doctrine of ethical limit is not a threat to democracy, it may indeed be the only thing that saves democracies from themselves.

The life of the soldier as it relates to the question of limiting obligations raises an intriguing paradox. On the one hand, we have a group of men who have willingly undertaken to live a life that demands special sacrifice and hardship and the subordination of their own interests to the pursuit of the common good. On the other hand, these men, who have taken an oath to bear the burden of unlimited liability in combat, are usually reticent to speak out against orders they find repugnant or to openly disagree with their superiors.

Perhaps one of the greatest mysteries of the military profession is the fact that so often the officer who is willing to sacrifice his life in combat is hesitant to risk his career to correct an abuse in the system, to suffer the embarrassment by speaking out for justice, or to stand firm on moral standards when the accepted practice follows a discordant tune. Being a brave combat leader does not guarantee that an officer will have the courage to overcome pressures to behave unethically in a bureaucracy. It all comes down to his personal standards of integrity and a sense of conviction for his service calling.[8]

The soldier must come to grips with the problem of the limits of military obligation, but he cannot do it by ignoring the problem. He will not solve the problem by going along with orders he feels are immoral. The only way the soldier can deal with the problems of dissent, loyalty, and resistance within the profession of arms is to address the question openly and to specify legitimate ethical alternatives to acquiescing in orders one regards as immoral.

Two basic questions relate directly to the limits of the soldier's obligation to obey orders. (1) What is the moral responsibility of the professional soldier when he is called upon to execute a policy to which he has a moral objection? The answer in principle is simple: no soldier can escape moral responsibility for his acts, and his obligations can never bind outside the context in which they must be observed. Accordingly, a soldier may not carry out any policy to which he has a genuine moral objection. (2) What are the limits of professional obligations when duty is confronted with conflicting moral imperatives? It may be answered that when moral imperatives conflict, judgment must be exercised relative to the circumstances in which one is required to reach a decision. From these two perspectives, the rest of the discussion on dissent, loyalty, and obedience will proceed.

Not all rules have an ethical content, and not all disagreements involve ethical issues, nor are all obligations moral ones. Therefore, the problem of limits addresses large moral dilemmas, and not trivial concerns. Such moral dilemmas generally do not arise very often within the normal context of the military profession, although they are likely to arise on the field of battle and as such emerge more often than in any other professions. Of course, a soldier may seriously disagree with an order and have good reasons for doing so but without necessarily involving an ethical conflict. The point is that while moral conflicts do not arise very frequently in the normal course of things, when they do they must be dealt with.

In both policy-making and execution, there is a normal area of disagreement, and men must often execute policies they think are less than ideal but do not involve any serious moral issues. When discussing the limits of loyalty, of obedience and, indeed, of the moral requirement to resist immoral orders, the focus is on those serious obligations that have a genuine moral dimension. When ethical questions are not involved, the soldier's rule of obedience is expected to apply. It must also be clear that the soldier is not being counseled to disobey or resist orders per se. This discussion explores the question of human action as it relates to members of the military when they are confronted with serious ethical problems.

The experience of the American military during the ten years of war in Vietnam and, indeed, even at times since then, revealed a number of ethical deficiencies that seemed almost endemic to the military organization. Among the most important of these is a military careerism and loss of ethical orientation so severe that the protection and advancement of an officer's career position at all levels has become the highest operant value for a substantial number of officers. This turning away from traditional military values has not been without its pragmatic effect on the operational capabilities of the military. In general, however, the change in the profession's value structure has resulted in a series of ethical failures defined in terms of officers acquiescing in, initiating, or participating in policies that are seen as unethical but that are followed anyway for lack of any clear ethical alternative or because of the rewards that redound to the soldier's career. The deliberate falsification of intelligence reports from hamlet to strategic level, body count doctrines, rules of expending ordinance, free-fire zones, and false readiness reports, to say nothing of the "deals" made to get the troops to reenlist, are instances of the problems. More important, the same type of unethical behavior is evident in the peacetime military largely because the forces that generated it then are very much with us still.

The Problem of Acquiescence

Why did this situation occur, and why does it persist? Why did the officer corps fail in its stated duty to make ethical choices? The exaggerated emphasis on career success which provokes unethical actions and acquiescence in questionable orders is the result of the profession's failure to develop ethical doctrines of dissent and resistance. Many soldiers are interested primarily in the advancement of their careers. Such advancement is often purchased at the expense of a failure to question the orders of superiors, regardless of their operational and ethical consequences. The result is that the Vietnam era witnessed the emergence of an officer corps whose members all too frequently acquiesced in policies, orders, and actions that many felt were wrong from the perspective of personal ethics but were executed anyway.[9]

The extent of the problem of ethical failure is obvious and embarrassing. In the ten years of war, not a single general officer resigned in protest over the policies pursued in Vietnam. Indeed, I cannot find a single instance where a general officer refused, by way of resignation, to execute a single policy, although in retrospect it appears that a substantial number of them did have serious ethical and pragmatic objections to the policies they were asked to execute.[10] If we can trust the postbellum statements of these high-ranking officers, they were opposed to such specific policies as search and destroy, body counts, ordnance expenditure doctrines, falsified intelligence reports, and the bombing of rural populations to force them into strategic hamlets. Whether they disagreed at the time they were executing these policies is an open question. What is undeniable, however, is that not a single general officer resigned in protest or otherwise gave voice to his ethical objections. Douglas Kinnard, in *The War Managers*, an extensive survey of general officers who served in Vietnam, found that many of them had serious ethical and practical misgivings about policies they carried out.[11] Yet, not a single officer resigned. Indeed, in the twenty-year period between 1960 and 1980, a period encompassing American involvement in Vietnam and a wide range of major policy debates including the B-1 bomber, Salt II, and the All-Volunteer Force, only one general officer, Major General George Rowney, resigned in public protest over policy and that over Salt II. During that same time period, twenty-seven officers of flag rank in the Canadian forces resigned in public protest over questions of policy. The contrast is stunning.

Despite evidence that specific policies had been failing for years and that a whole range of policies implemented between 1972 and 1980 were also failing and provoking public debate outside the military, not only did no resignations occur (except for General Rowney) but also there is little evidence that any senior officers seriously protested even within the establishment. An equally disturbing fact is that with the single exception of Colonel David Hackworth, a much decorated war hero, not a single colonel or even a lieutenant colonel resigned in public protest. The only clear examples of resignation in protest occurred at the lower rank levels of the officer corps and even then only rarely. The propensity is to keep one's mouth shut and to go quietly over the side, saluting as one steps into the ethical abyss. Of course, a few officers, largely during Vietnam, sporadically refused to execute orders, but in almost all cases such actions were confined to junior officers who were OCS or ROTC graduates and did not intend to make the military a career.

Resignation in protest, resistance, and even criticism as ethical alternatives to acquiescing in orders were not established practices during Vietnam, or during the period immediately following the war, and they are not established practices now. The sad truth is that the military has been unable to develop and institutionalize a doctrine or code regarding a soldier's obligations with respect to the questions of loyalty, dissent, and even disobedience. When confronted with a policy he regards as unethical, the soldier is usually unaware of what his obligations are. In short, there has been no formal recognition of or training in what may be called the legitimate avenues of military protest that can operate in a democratic society.

Legitimate Avenues of Military Protest

An awareness of his obligations and the available avenues of ethical protests makes it easier for the soldier to exercise moral options. The fact is that one cannot rely on the soldier's virtue, for even the most virtuous soldier may not be aware of his obligations or know how to act. An ethics of virtue is important, but by itself will not serve as a substitute for a clear statement of obligations. In this view, the argument that one can rely completely on the integrity of the soldier with little concern for making the soldier aware of his obligations or the procedures for carrying them out is not valid.

If one creates procedures that allow the expression of dissent and even disobedience within the military, the bureaucracy itself may co-opt these procedures, especially those associated with resignation, and render them empty and sterile gestures. Under these conditions, statements of moral protest, especially resignation in protest, may be reduced to bureaucratic exercises and rob them of their content as examples of moral action. The amount of dissent and resistance to immoral orders may even decrease.[12] There is always the risk that the organizational bureaucracy will neutralize procedure for dissent. That does not mean, however, that the procedures have no inherent value. Moreover, it is the obligation of the profession's leadership to insure that the bureaucracy does not pervent noble ethical acts by reducing them to paper exercises. In any case, it must be remembered that resignation, dissent, or protest are likely to occur only rarely. Thus, the bureaucracy could not realistically develop and sustain the kinds of practices that could corrupt and pervert them. But even if it could, the members of the profession still have an obligation not to allow the seriousness of an ethical act to be turned into a bureaucratic procedure and thus be nullified.

An alternative argument against any form of dissent, resignation, and protest is the position put forward in defense of the soldier's failure to recognize moral responsibility. The argument suggests that it does no good to get mired down in such questions to begin with. If an officer were to resign his position, dissent, or protest, it would probably ruin his career, isolate him among his peers, or result in his dismissal from command or the profession itself. The organization would simply find someone else to do its bidding, and the game would go on. Moreover, the possibility always exists that one's replacement would be even less sensitive to ethical concerns.

The difficulty with this position is that it completely ignores the point of moral obligations. Responsibility for observing moral obligations remains whether or not observing the moral obligation does any good in changing policy. One still has the obligation to try. No one would seriously argue, for example, that the Ten Commandments should be ignored as rules of behavior because people violate them. Moral responsibility remains regardless of the degree to which it is observed. If the question is asked as to who benefits from ethical action if the organization ignores it, the answer is that it benefits the individual who kept faith with his sense of ethics. The fundamental point is that sometimes the value of ethical action resides with the individuals who observe it, individuals who can never escape their ethical responsibility. From this perspective, their obligation to be true to themselves and their ethical code transcends the degree to which it is effective in changing the policy of the organization. It is simply not important whether the bureaucracy responds.

Nonetheless, sometimes the bureaucracy does respond, and that possibility always exists at least in principle. In Canada, as noted earlier, during 1960-1980 twenty-seven flag rank officers publicly resigned in ethical protest over questions of policy. In Canada's case, such resignations did reverse policy. Yet, even if no policy changes occur, the fact that individual officers resist policies to which they are ethically opposed or resign in protest or refuse to execute immoral orders breathes life into the meaning of the profession. It creates moral exemplars, role-models, and, in some instances, even the stuff of legend to whom young officers can look for guidance. The creation of ethical examples of excellence further sustains the meaning and integrity of the profession's code of ethics. It establishes precedents from which other officers and soldiers can learn. In the final analysis, the fact that an ethical act may or may not do any good in a bureaucratic sense is not a particularly relevant criterion for judgment. If one looks to such men as Socrates, Thomas Beckett, and Thomas Moore, it is clear that their actions stand by themselves, although their actions did not have immediate impact on the official policies of the organizations they tried to influence. Yet, they remain moral exemplars, their actions breathe life into ethics, and their actions have ethical value.

If the military profession is ever to recover from the debacle of Vietnam, as well as the policies promulgated since the end of that war, it must first undergo an ethical rebuilding as an essential precondition for its further operational rebuilding. Among its first priorities must be to develop a code of military ethics along the lines suggested in Chapter 5. It must also develop among its officers and men a capacity to balance ethical and career pressures. It must develop within its officers and men a capacity for moral reasoning, ethical judgment, and the personal courage and institutional support to exercise moral options. This capacity is required at all rank levels but especially in the officer corps where the ears of policy-makers are most readily accessible. It is imperative, then, that the military develop a doctrine of ethical protest for its members and that it support the exercise of this doctrine in pragmatic terms. *Of course, this doctrine must be consistent in theory and in practice with the values of democratic society and continued civilian control of the military apparatus.* Any doctrine violating these basic precepts would be unacceptable and dangerous, tending to provide justification for excessive military influence within a democratic civil order. That is not the intention of any honorable soldier serving within a democracy.

What, then, are the ethically permissible avenues of protest for the soldier consistent with the democratic values of a free society and continued civilian control of the military? What courses of action may a soldier properly take when faced with the problem of being ordered to execute or acquiesce in policies and orders to which he has moral objections? Four legitimate avenues of military protest are open, and all are consistent with the basic precepts already discussed. They are (1) resignation or retirement in public protest; (2) request for relief in protest; (3) appealing of orders to a higher command; and (4) direct refusal to execute an order. None of these alternatives conflicts with the democratic persuasion, and all are congruent with the Western military tradition. Since none is inherently associated with collective resistance, the menace of the coup d'etat cannot be associated with them. It is important to observe that the American military has never developed a doctrine of moral guidelines for resistance to immoral or ethically unacceptable orders. It is precisely in this official moral vacuum that the exaggerated values of careerism and self-interest operated with such force in Vietnam and continue to operate so forcefully today. There is a need to reduce that moral vacuum.

RESIGNATION/RETIREMENT IN PUBLIC PROTEST

The most obvious way in which an officer can demonstrate his disagreement with or moral outrage toward a policy is to resign from the profession as an act of public protest. He must leave the service and his profession, and seek to influence his government from the outside as a citizen. This is his legal right and, in some instances, even his moral obligation. Accordingly, resignation or early retirement as a response to ethical pressures implies that there can be no escape from one's ethical obligations; at times, that obligation continues even after one has left service. If the precepts of the profession are no longer observable because they require the sacrifice of an individual's sense of ethics, or if they run counter to his sense of obligation to a higher sense of morality, or, as is most likely, if extant practices run counter to the profession's own stated code of ethics, the soldier must retire or resign. He puts himself beyond the community by his act of resignation, and he is then no longer bound by the precepts and limits of the community. In such an instance, he may feel free, and even obligated, to try to change or stop policies he regards as ethically offensive but must now do so in his role as a citizen.

Resignation can be accompanied by a public declaration of the reasons compelling an officer to resign, thus exposing the policy or orders in question to public scrutiny and debate. Such a course of action is consistent with democratic values and in no manner challenges civilian control of the military. In a practical way, resignation or retirement in protest presents evidence to the system itself that its policies may be in serious error or causing consequences that were not foreseen. To that extent, dissent will increase the rationality of the decision-making process by increasing the amount of information available to it.

Resignation or retirement in public protest is almost always a more powerful resource when used by a general or other high-ranking officer. Indeed, it is likely to be the most powerful means that a general officer can employ to effect a change in policy by focusing attention on the objectionable policy itself. Since he is likely to be closer to the decision-making level than his subordinates, the resignation of a general officer can be expected to have greater public impact. While the degree of impact that the resignation of any officer will have on any given policy is always an open quesion, in the main it seems rather obvious that no unethical policy will be easily changed from within the bureaucracy if powerful bureaucratic elites have a vested interest in its perpetuation. The threat, then, is to go outside the system in search of change. The threat and willingness of military professionals to publicly resign in protest to policies they regard as ill advised, not in the best interests of the country, or blatantly immoral is a legitimate way of carrying out their ethical obligations to the profession itself. Public resignation may force civilian policy formulators to make policy with a view toward how the military would publicly react to it. In this sense, a kind of law of anticipated reactions may begin to operate in that civilian policy-makers are likely to be more sensitive to some of the harsher aspects of military policy if they have to defend them in a public debate provoked by the resignation of a military professional whose stock in trade is his expertise in the policy area being debated. Thus, the officer who resigns acts as a lightning rod to draw the attention of both his military and civilian superiors, as well as the concerned citizenry, to morally objectionable policies. At the very least, it will force his superiors to defend those policies.

Pragmatically, of course, the general officer risks little in the way of earned career benefits by his resignation or retirement since in most instances he will take his benefits with him. This is not the case with lower ranking officers who may find the price of public protest to be an end to their military careers. Thus, when confronted with an order or policy that is morally objectionable, the general officer easily has the best chance of making his objections felt. He has the ears of high-ranking policy-makers and is identified in the public mind as a figure of some importance as he is among the soldiery. In any case, he relinquishes only terminal career goals. He will certainly fail to make the next promotion list and will never be promoted to chief of staff. Yet, given the excellent record of public resignation over ethical questions that one finds among flag officers of the Canadian military, there is evidence that truly professional offices are willing to pay the price.

The question of resignation or retirement in protest has graver dimensions for the soldier than for members of other professions. The soldier has a more serious problem because his employer is the state itself which has a complete monopoly on the employment of his skills. Unlike doctors and lawyers who may resign from one practice and start another practice elsewhere, soldiers have no such option. They may only legitimately apply their skills in the service of the state, and their resignation or retirement quite absolutely puts them beyond their profession and there is no going back. To require that military men respond to unethical orders by resignation or retirement in protest is to place yet another heavy burden on men who already carry much heavier burdens than their civilian brothers. Nonetheless, the only alternative is to permit the soldier to escape responsibility for the execution of orders that he regards as immoral or unethical. This course of action can never be tolerable in a society that claims to have ethical limits nor, indeed, tolerable in a profession that claims as its core ethic the quality of selfless service. Thus, the soldier in one more respect finds himself in a profession that is different from other occupations and professions. That is why the price of belonging is high and why not all men are fit for the profession.

Drawing upon the doctrine of *respondeat superior*, it may be argued that high-ranking officers, especially general officers, have a greater moral obligation to protest unethical actions or orders because their position carries with it a stronger moral obligation to seek the welfare of their subordinates and the country they have sworn to serve. In the American military, high-ranking officers generally tend not to speak out against policies or orders of superiors. Few of them are apparently concerned about the issue at all or are able to frame their disagreements over policy in ethical terms. While all members of the military carry essentially the same ethical obligations and are responsible for the same actions, the higher one's position in the organizational hierarchy, the greater the scope of one's responsibility, inasmuch as the damage one can cause by failing to act ethically or by complying with an immoral order is far greater than for lower positions. For example, the horror caused at My Lai by a second lieutenant does not approach the scope of moral terror that can be wrought by a division commander who has no moral restraints. The higher ranks of the military, those who are directly responsible for advising the president and Congress as to what is best for the country, do indeed have awesome ethical responsibilities. Accordingly, they must be prepared to defend their professional judgment and professional ethics by overt acts of dissent when necessary and, if circumstances warrant, even through the act of public resignation or retirement in protest.

Resignation or retirement in protest, while an appropriate course of ethical action, cannot be expected to occur in large numbers among the junior officer corps except in extreme cases. Junior officers cannot realistically be expected to abandon their livelihood and careers in midterm except under extreme moral pressures. More importantly, the resignation of junior officers, unless done in mass, is likely to have only limited, if any, effect on policy. Yet, American history contains a number of examples, especially during the Indian Wars and the Philippine War, of young officers prepared to resign on moral grounds. It is not suggested here that the failure of junior officers to resign in protest in deference to perpetuating their careers is an admirable act, only that it is understandable. This is precisely why alternatives to resignation or retirement in protest consistent with ethical action must be found. Otherwise, the junior officer corps or young soldiers have little chance of making a moral choice in the face of immoral orders. Required are additional avenues of ethical action for the soldier that are available to the majority of soldiers below the general officer level.

REQUEST FOR RELIEF IN PROTEST

To argue that resignation or retirement in protest will not be very common or, perhaps, a very effective organizational alternative to present conditions does not imply that junior officers or common soldiers are relieved from their obligation to take action in the face of policies or orders regarded as immoral. Obligations, as noted earlier, spring from many sources, not the least of which are the values of the society that the soldier has sworn to uphold as well as from the profession of which he is a part. Such obligations exist whether or not the means to carry them out are functional to career success. The case for situational ethics is rejected here, as it was rejected earlier.

However, if ethical choices are to be made with any kind of regularity, the soldier must be provided with reasonable options to make an ethical choice when he confronts orders he regards as immoral. When confronted with such "local" policies as shooting prisoners or civilians, burning civilian dwellings, poisoning wells, or murdering children, the soldier has the legitimate right to formally request that he be relieved of his duties to participate in such actions. His request should detail the policies themselves as well as the reasons why he thinks they are immoral, illegal, or unethical. This action, taking place in writing as well as orally, immediately engages the military bureaucracy and has the important effect of creating a written and oral record of events as well as bringing the events in question to the attention of higher command and staff. Even the most hidebound of bureaucracies finds it difficult to ignore written records.

By filing a written and oral report in the form of a request that one be relieved from duties on the grounds that the duties themselves are illegitimate, the soldier has a very practical way to discharge his obligation to act ethically. He also observes an obligation to his profession by making known improper orders or policies to his superiors. At the very least, this course of action reduces the possibility that superiors can hide behind the doctrine of "plausible denial" by claiming that they did not know what was happening within their areas of responsibility. Plausible denial was the implicit basis of the defense pleaded by Captain Medina and ultimately by Major General Samuel Koster to avoid implication in the My Lai affair. Plausible denial violates both the precedent of *In Re Yamashita* and the historical responsibilities inescapably linked to all commanders under the doctrine of *respondeat superior*. For example, Yamashita was hanged not because he ordered the atrocities of the Bataan Death March; indeed, the evidence was that General Yamashita as commander had little knowledge of the conduct of his forces at the time and even less control. He was hanged because of the clear application of the historical military principle that a commander is ultimately responsible for the immoral acts of his command. This principle has long applied to Western military custom and practice. Thus, the request for relief in protest focuses responsibility on command elements responsible for issuing immoral orders. As such, it becomes a legitimate avenue of ethical choice for the soldier confronted with the problem of having to execute such orders.

To be sure, not all requests for relief or transfer will be granted. Yet, if the issue raised is one of illegality or immorality, the request will not likely be blocked at the lower levels of the bureaucracy. These levels will most probably seek to avoid what promises to be a very difficult decision and tend to transmit the report rapidly up the chain of command. Indeed, even at My Lai the field reports of the massacre were transmitted rapidly up the chain of command until blocked by the division commander to conceal his own complicity. A request for transfer or relief on the grounds that localized policies or orders are unethical or illegal provides a soldier with a viable and practical means for exercising his moral obligation to the profession's code of ethics consistent with his position within the military hierarchy. There will certainly be costs, especially to the young officer who seeks a career, and especially if one's superiors ultimately decide that what the soldier did was incorrect despite his intentions. Nonetheless, the obligation to object to unethical policies, orders, and practices remains. The request for relief or transfer represents a legitimate way within a democratic society for a member of the military profession to carry out a moral obligation. No one ever said it would be easy, only that it would be possible and necessary.

APPEAL ORDERS TO HIGHER COMMAND

The assumption in any military organization, especially one as dedicated to democratic values as ours, is that illegal, unethical or immoral orders will not be deliberately issued *as a matter of official policy*. Local commanders may overtly or covertly condone or encourage such policies, but they are decidely local policies and do not represent the official policy of the profession or the state it serves. This distinction opens up yet another avenue that a soldier may choose in maintaining a moral position when confronted with the dilemma of carrying out orders he regards as unethical. The soldier may legitimately take the step of "going over the head" of his superior as a formal means of protest. Whereas the object of this alternative is to bring to the attention of higher authorities the orders and policies that are morally objectionable to the soldier in the hope that they will be changed or stopped, the directed charge is that the immediate ordering commander is exceeding his authority by formulating policies and ordering activities that his superiors, both military and civilian, would not permit if they knew about them. The assumption underlying this course of action is that the profession remains a repository of ethical trust and that those who hold that trust in their charge can be counted upon to uphold it when brought face to face with evidence that someone is violating it.

To some extent, the military does provide for this alternative through the office of the inspector general. Those familiar with the operation of this office are aware that protests handled in this way are often officious, bureaucratic, and slow, to say nothing of provoking feelings of disloyalty and penalizing the soldier who had the temerity to initiate the investigation. The recommendation is that the concerned soldier go to the relevant superior within the chain of command. Conventional military wisdom holds that remedial action will occur fairly rapidly, a fact that can be of grave importance if the orders involved deal with the shooting of civilians and the torturing of prisoners. Furthermore, such an appeal makes it clear to all those positioned at different levels within the chain of command that something is amiss. They may not respond, but their silence will be more difficult to purchase even by their immediate superiors who may be involved in unethical acts. Yet, some evidence suggests that even these remedies may be neutralized by the compulsion to "team playing." Even so, a determined soldier or officer does have an avenue of redress if he chooses to pay the career price. Indeed, he may not legitimately shrink from paying that price without accepting or acquiescing in orders he regards as wrong. Confronted with this Hobson's choice, the directions of ethical action are clear: the soldier must refuse to obey, or he must make every effort to bring to the attention of his superiors or other appropriate authorities the fact that he is being ordered to do something he regards as wrong. If he refuses to act, he betrays his oath and his profession.

REFUSAL

The stress in this chapter on the ease or difficulty with which a given course of ethical action may or may not be implemented should not obscure a basic point: a moral obligation not discharged in the face of surmountable practical difficulties remains no less an obligation, and the soldier remains responsible for his failure to observe it. Inescapably, all members of the military will encounter a situation in which they attempt to change an order or policy in other ways, or in which the practical cost of implementing their moral obligations may be ruinous to their careers. Difficult as these circumstances might be, the soldier's obligation to ethics remains.

In any situation of obligation and obedience, the ultimate response the soldier can make to orders requiring actions he considers unethical is to refuse to carry them out. The course of refusal is the last resort and the response to extreme ethical pressures. Moreover, it is premised on the assumption that the soldier is willing to accept the consequences of his act *if it is later judged to have been wrong.* An important point is that the refusal to execute an order is not the end of the judgmental process of ethical assessment. Whether in military or civilian life, an act of refusal is judged at a later time as to its acceptability or unacceptability. The willingness to "accept the consequences" of one's act of refusal is really a statement of readiness to justify one's action at some appropriate time. It is not an assumption of *a priori* guilt or of prepardness to accept summary justice on the spot.

Even a doctrine that repudiates disobedience in the large, as indeed the code of ethics put forth here does, must leave room for the refusal of obedience in particular cases. A soldier faced with conflicting ethical obligations must choose one over the other when he cannot do both. He may easily find that disobedience and the refusal to execute an order are the only ethical paths open to him.

Still, some would argue that the soldier should be loyal to his superiors or to the state regardless of the nature of their orders on the grounds that the soldier is only the technical instrument of the will of the state. If any one is to be held responsible for the actions of the soldier, let it be his civilian or military superiors and let the judgment be made by the victor. Like the sense of judgment attendant to the theory of the divine right of kings, only history (or God) may judge man's actions. No amount of technical expertise or attempt to reduce the soldier to a mere instrument of someone else's will can ever be a valid ethical doctrine within the historical and value context of the Western world. In a serious ethical crisis involving superiors, a subordinate must never confuse his loyalty to his oath, the Constitution, and the profession's ethics with loyalty to his superiors as persons, even civilian superiors. General Douglas MacArthur, while himself involved in an ethical crisis concerning his proper role as a subordinate responding to what he perceived to be higher obligations, expounded a valid ethical position with regard to the loyalty of the soldier trapped in an ethical press.

I find in existence a new and heretofore unknown and dangerous concept that the members of our armed forces owe primary allegiance or loyalty to those who temporarily exercise the authority of the executive branch of government rather than to the country and its Constitution which they are sworn to defend. No proposition could be more wrong or more dangerous.[13]

In short, a soldier's moral obligations transcend and surpass the obligations owed to his immediate superiors and even his civilian superiors in certain condtions. General Marshall, the epitome of the loyal soldier, was echoing MacArthur's sentiments when he said that "an officer's ultimate commanding loyalty at all times is to his country and not to his service or his superiors." In a crisis, the soldier must exercise his sense of loyalty as *fides*, and it must always take precedence over any sense of *obsequium*. Indeed, the problem is even more complex, for in a deep moral crisis the soldier may even have to override his oath to the profession and to the Constitution in order to be loyal to humanity itself.

The Germans, who perhaps have had more direct experience with officers and soldiers being crushed between demands of their oath and the course of immoral events, have developed an interesting distinction in dealing with the question of loyalty to superiors. They distinguish between *hochverrat* and *landesverrat*. *Hochverrat* is disloyalty to a superior, which in Germanic terms meant disloyalty to the monarch or other governmental head of state. *Landesverrat*, by contrast, is disloyalty or betrayal of the nation. Within this distinction there is room for maneuver in making an ethical choice. In order to serve the nation or the Constitution, a soldier may sometimes have to be disloyal to his superiors or refuse to execute their orders. The Germanic distinction between the two notions of loyalty throws into focus what every member of the military profession knows in his heart, and that is that fundamentally a soldier's first loyalty is to behave ethically and humanly, and that in times of severe moral crisis he must be prepared to follow that higher morality.

Some reject this line of reasoning on the basis that it erodes the sense of duty that is fundamental to the military. The difficulty with this argument is that it uses the concepts of loyalty and duty far too loosely. When speaking about "doing ones's duty," it must be clear that to be dutiful requires a sense of being bound only by what is ethical and moral—it is to bind the soldier in the context of the ethics of his profession as to what is acceptable behavior. As with all ethics, obligations apply in circumstances, and the obligation to do one's duty also applies in circumstances. The proper function of duty is to make the soldier sensitive to the relationships and claims made upon him in particular situations, so that he knows what his duty is, that it arises in a social context, and why he has a duty to obey certain obligations. In essence, to be an ethical soldier is to do one's duty as to what is ethically right and to know why those ethics bind. Duty is not to be blindly tied to following orders.

When duty is tied to a knowledge of the reasons why one is bound, one has an ethical obligation to do one's duty. But to do one's "duty for duty's sake" is a perversion of its true meaning simply because it isolates the actor from the very reasons why he is required to act. Duty no longer is the servant of the claims made upon him by reason of his membership in the profession, that is, it is no longer the obligation to observe its ethical precepts. Instead, duty becomes a replacement for these precepts. That is precisely why to execute immoral orders is not "doing one's duty" in the proper ethical sense of the word. The duty to serve a code of ethics is based on judgments about its moral applicability in a given context. To do one's duty when the application of the code does not tend to achieve what the code intends is wrong; it is also no defense. Thus, the argument that soldiers who are allowed to refuse to execute orders they judge to be immoral will not do their duty is based on an erroneous conception of the notion of duty. The duty of the military professional is to do what is ethically right; one can never have an ethical duty to do that which is immoral. To interpret duty as the requirement that the soldier carry out all orders of his superiors simply because they are orders is to misunderstand both the concepts of duty and ethics.

Yet, the concept of duty is frequently misunderstood within the military. Joseph Ellis and Robert Moore discuss the problem of misunderstanding what duty implies for the soldier. While their comments apply directly to West Point cadets, they may be extended with equal veracity to a large number of officers within the military itself.

When caught in a moral dilemma, most West Pointers are conditioned to perceive their obedience to lawful superiors as the highest form of duty. Such a perception is regarded as the essence of military professionalism, for it involves putting personal considerations beneath service, duty to one's self. When there is a conflict between what a West Pointer calls duty and honor, then, he is likely to have no ethical answers. Or rather, he is trained to answer by equating honor with duty.[14]

This condition represents an ethical failure of the first magnitude as it relates to the soldier's obligation to refuse to carry out orders he regards as unethical. It equates obedience with obligation and obedience with honor. Obedience is simply carrying out orders without comprehending why one has to do so; obedience may even legitimately involve an element of duress. Obligations, once again, require the willing execution of legitimate orders and require that one comprehend the reasons for them. Moreover, the condition portrayed at West Point confuses duty with compliance. It hints that by equating honor with duty, by submitting one's will to that of a superior, that a soldier can escape responsibility for his acts. This view is not only ethically wrong, but, as I have tried to show, also legally improper within the context of Western law. Finally, it implies that an officer can under some conditions suspend his ethicial judgment with impunity. Nothing could be further from the truth. Ethical responsibility is rooted not only in codes of professional ethics but also in Western notions of law. In neither can one suspend one's ethical judgment without penalty.

All of the conditions noted in the description of cadet behavior at West Point seem to underscore exactly what is wrong with much of the military profession's understanding of ethical obligation, especially as it relates to the problem of immoral orders. The refusal to execute an order is a comment not so much on the refusal but on the nature of the order given. In any event, one can never have an obligation to do what one ought not to do, and no soldier ever ought to execute orders he feels are immoral. To suggest that there is an escape from this dilemma by affirming that the soldier ought to suspend his ethical sense or submit it to that of another in deference to discipline, the mission, loyalty, duty, or any of a hundred possible reasons is to counsel further unethical action. It is also a doctrine that destroys the military's attempts to maintain an ethical center.

The refusal to carry out an order issued by a legitimate authority is prima facie an illegal act, although not an immoral one. Furthermore, refusal to obey is a way to make a moral choice not only immediately by the singular act of disobedience, but also in another way. Any refusal to obey an order immediately engages the military's legal conflict resolution structure, namely, the court martial, in much the same way that the technical violation of civil law is necessary to engage the civil courts. The courts then become the mechanism for having the order or law that was disobeyed judged definitively in terms of its application. The engagement of the court martial system because of a soldier's refusal to execute an order he believes to be immoral or illegal provides two opportunities. First, it provides a public forum in which the soldier may fully state his moral case in an attempt to justify his action. Second, it provides the military with the opportunity to evaluate the circumstances of the incident relative to the order issued and to take appropriate action against the issuing authority if justified. Thus, the military court system is a two-way street. Like the American legal structure, it can only respond to a justiciable issue, and a justiciable issue can only be considered so *after* some order or directive has in fact been violated or someone has refused to execute it. Accordingly, the act of refusal to execute an order deemed to be immoral or illegal by a soldier constitutes an appeal within the military legal system to higher authority for a judgment on the original order itself.

From this perspective, refusal to execute an order on moral grounds becomes the military equivalent of technical civil disobedience in the service of a higher cause, the ethical sense of the profession itself. It is not the equivalent of disloyalty or cowardice in any *a priori* sense. The important point is that the refusal to execute an order is not the end of the judgmental process. Whether in military or civilian society, that act is judged at a later time as to its acceptability or unacceptability. Thus, the willingness to accept the consequences of one's act is a statement of one's readiness to justify one's actions at some appropriate time.

Obstacles

Some avenues of ethical protest are more practical than others, and some carry greater risks. Yet, the greatest risk of all for the professsion would be the inability of its members to take any ethical risks in its service. Even so, all the avenues of protest outlined here are legitimate in that they are consistent with the dominant values of a democratic polity that the soldier serves. Furthermore, they are consistent with the dominant values inherent in the profession of arms, namely, the requirement that soldiers in the performance of their task are still limited by the boundaries of a code of ethics which serves to mitigate the harsher and more terrible aspects of collective violence. A major ethical obligation of any soldier is to ensure that his conduct as well as that of his superiors is generally consonant with the ethics of the profession. None of the courses of action available to effect moral protest within the military may be construed as an ethical basis for massive disobedience of civilian authority by the soldiery. There is no question of a coup d'etat.

What is interesting in attempting to deal with the problem of moral protest within a military environment is the fact that the principal democratic nation in Western society, the United States, has been unsuccessful in promulgating a doctrine concerning the subject, while other nations have developed an ethos of protest within their respective military structures. For example, both British and French societies have long recognized the officer's right to resign in protest, and, indeed, he is expected to resign over questions of honor. Ironically, such a tradition may have developed from the fact that the military in both countries has drawn heavily upon the aristocracy for its members, at least in its officer ranks. This suggests that officers who chose resignation still had alternative, secure social and career roles to enter outside the profession of arms. In addition, since both Britain and France have long traditions of parliamentary government where resignations of heads of state and cabinet members are commonplace, it is possible that much of the trauma which we Americans associate with resignation from public positions is not present there. Canada is another example which has already been mentioned. The sterling record of the Canadian officer corps over the last twenty years, especially its flag ranks, in providing exemplars of ethical action is perhaps the best in the Western world. With respect to the Germans, mechanisms of moral protest have been preserved in the operation of its boards of honor and, in some extreme cases, in the legitimization of suicide as a permissible course of action. The commander of the Graf Spee and the suicide of Erwin Rommel are cases in point. The Japanese code of Bushido which required suicide from a soldier who felt himself in moral disagreement with orders is too well known to require further elaboration here. The number of officers, many high ranking, who took their own lives, suggests if nothing else that it was not an empty code. Now certainly some of these measures of ethical action are rather extreme, while others, such as the French, British, and Canadian examples, are entirely consistent with democratic values. The point is that military professions in other countries have been able to develop functional doctrines of resistance for the soldier caught between the demands of conscience and the orders of his superiors. The task, therefore, is at least possible.

If it can be assumed for the moment that many of the problems which I have described here are at least traceable to the military's failure to develop formal mechanisms through which soldiers can make ethical choices when confronted with ethical dilemmas of some magnitude, the point must be made that this failure to develop a *formal* doctrine of moral protest is only part of the difficulty. The fact is that formal rules in any bureaucracy will affect behavior only to the degree that they are supported and reinforced by the informal norms and values of the institution. The ethical failures of the officer corps during Vietnam were possible not only because the military lacked a formal doctrine of moral protest, but also because the informal rules of the military subsociety would have effectively undercut the operation of any such doctrine. The doctrines functional to career success, although informally articulated, often stand in stark opposition to those precepts requiring the soldier to make ethical choices. Stated otherwise, violation of the military's own sense of honor often pays off in career terms and in some instances is demanded by the terms of the system itself.

This tension between informal norms functional to career advancement and an attempt to develop a formalized doctrine of ethical protest within the military profession can be expected to persist despite the most sincere efforts at reform. Even with a clear code of ethical protest, informal norms have to develop and be deeply embedded in the soul of the profession, to the extent that the soldier who exercises his ethical prerogatives is not degraded or vilified by his peers or superiors. Members of the profession at all levels must come to see the exercise of ethical obligations as the highest form of loyalty to the profession. Furthermore, the actions of ethical men must also be functional from the point of view of career advancement in order to encourage their undertaking. As Lieutenant Colonel Harry Summers has said:

We temporize and apologize for those who violate our standards rather than rising up in outrage and indignation and casting them out with the scorn and opprobrium they deserve.... The Army can, and should, ensure for we lesser mortals that integrity, character, moral convictions, tenacity and fighting ability pay.[15]

At present, the soldier who goes over his commander's head, resigns from the service, or reports an unethical practice is commonly viewed as "disloyal" or a "quitter." The military has made individual loyalty an absolute, while almost ignoring or even neutralizing larger loyalty in the form of an ethical commitment that the profession must have if it is to engender in its members a sense of communal worth. As long as the principal stress remains upon individual loyalty to superiors and as long as violations of ethics are allowed to pay in terms of career enhancement, it is unlikely that any formalized doctrine of ethical protest will be adopted. Yet, the stakes are too high not to try.

The case for a formalized doctrine of ethical protest within the military is not without its opponents, especially as it addresses the twin precepts of resignation and refusal to execute orders. The argument against developing and implementing such a doctrine is that if every commander was forced to explain every order (or at least demonstrate that it was not unethical) to every soldier and officer in order to gain compliance, the military would border on paralysis and there would be serious question if it could effectively carry out its mission. The argument is not overly convincing. Generally, questions of moral choice do not arise so frequently as to merit the charge that all or even most orders would have to be justified to subordinates in advance. Indeed, if conditions provoke a substantial number of officers and soldiers to demand such justifications, this in itself indicates that the military structure has already approached breakdown. If a large number of officers and men were forced by their ethics to conclude that a large-scale questioning of certain policies and orders is needed, we would merely be witnessing the symptoms of a disease that in all probability was already terminal.

Resignation as a course of action to register ethical protest is most often criticized on the grounds that it amounts to "quitting." Why not, the argument goes, stay within the system and work to bring about change? To the extent that the argument has any merit, it is more applicable at the general officer level where access to policy-makers is possible and advice may be heeded. While the question of future success remains open, the available evidence suggests that working from within the system does not usually work, nor will it necessarily satisfy the immediate ethical requirement that one may have to do something about an unethical policy now. Furthermore, an ethical decision delayed is one condoned. A long career of ignoring or condoning unethical acts, despite the intention to change things once a person gets to the top, may in fact ethically deform a soldier to the point that he becomes incapable of resisting later on. Finally, working within the system deprives the rest of the profession of moral exemplars. We have had too few in our military in recent years and we sorely need more. In the end, the decision to defer an ethical judgment on the grounds that it will be changed later does not relieve the soldier of his ethical responsibility at the moment. It seems more probable that the system is much better at changing the dissenters than the dissenters are at changing the system.

Conclusions

One major effect of the military's failure to develop doctrines of ethical protest legitimated for use by its officers and soldiers is the tendency for values that are functional to career advancement to take precedence over or to act as substitutes for ethical judgments in the face of questionable orders and policies. Accordingly, careerism runs rampant, and no dissent is heard—all of which constitutes a danger to truly effective military operations. It provokes the worst type of disloyalty under the guise of loyalty, namely, a marked failure to question policies and orders that often do not work or, that extract too high an ethical and moral price for their success. What is needed is the development of a formal doctrine that teaches officers and men the accepted avenues of ethical protest open to the soldier and encourages them, through the support of informal organizational values, to travel these avenues without fear when urged to do so by the press of professional ethics and personal virtue.

The fact that other military professions in other cultures have developed such doctrines to serve the same ends in their armies is proof that the task is not impossible. We must always take care to ensure that the pathways of moral protest for the military remain consistent with the democratic values of the polity as a whole and are never permitted to become an excuse for coordinated military action against properly constituted and acting civilian authority. In the end, we have far less to fear from a military force full of ethical men than we do from one full of careerist values and self-seeking entrepreneurs. Such a military profession can only become increasingly out of touch with the values of the democratic civil order that it serves, as is even now evident in the social composition of the All-Volunteer Force. Furthermore, it risks disaster for the civil order and itself, not through design, but through incompetence as manifested in its increasing inability to challenge and resist policies that run contrary to its professional judgment. The refusal to act ethically, to dissent, resign, or refuse to be a part of unethical or harmful policies as a member of an ethical profession, is a serious failure of ethics. It is also an act of moral cowardice.

Notes

1. At least this is the evidence as presented by Douglas Kinnard, *The War Managers* (Hanover, N.H.: University Press of New England), 1977.
2. This argument, already cited earlier, is found in Reginald J. Brown, "The Meaning of Professionalism," *American Behavioral Scientist* 19, No. 5 (May-June 1976):511-522.
3. Arthur Dyck, "Ethical Bases of the Military Profession," *Parameters* 10, No. 1 (March 1980):43.
4. Quoted in James Toner, "Sisyphus as a Soldier: Ethics, Exigencies, and the American Military," *Parameters* 7, No. 4 (1977):6.
5. Ibid.
6. This argument is found in the AWAC Study, the Drisko Study, and the piece done by Ayers and Clement, all cited earlier. It is also found in Josiah Bunting, "The Conscience of a Soldier," *World View* 16, No. 12 (December 1973):6-11.
7. Bunting, "Conscience of a Soldier," p. 10.
8. Francis A. Galligan, *Military Professionalism and Ethics* (Newport, R.I.: Naval War College, 1979), p. 78.
9. This argument is found in Richard A. Gabriel and Paul L. Savage, *Crisis in Command: Mismanagement in the Army* (New York: Hill & Wang, 1978).
10. Kinnard, *The War Managers*. This is, of course, the main thesis of his book.
11. Ibid., p. 75.
12. The danger that professional mechanisms of control will be co-opted and replaced by bureaucratic mechanisms is discussed in Dennis J. Palumbo and Richard A. Styskal, "Professionalism and the Receptivity to Change," *American Journal of Political Science* 18, No. 2 (May 1974):387-388.

13. Quoted in Zeb B. Bradford, Jr., "Duty, Honor, Country Vs. Moral Conviction," *Army* (September 1968):43.

14. Joseph Ellis and Robert Moore, *School for Soldiers: West Point and the Profession of Arms* (New York: Oxford University Press, 1974), p. 180.

15. Quoted from a book review by Harry G. Summers, *Military Review* (July 1972):108.

8 INSTILLING MILITARY ETHICS

The central problem of utilizing a code of military ethics eventually resolves itself into the question of the extent to which that code can be instilled in the numbers of the profession itself. To make a code of ethics effective within the profession of arms, the profession must make a serious effort to inculcate and instill its ethical sense and obligations in its membership. If the profession itself does not take steps to inculcate what it regards to be the core obligations that make the profession what it is, it will cease to be a profession. Worse, it will become a caricature of a profession in which its members lack a sense of special obligation, pursuing at whatever cost those goals that benefit their self-interest. Thus, the military and not the civilian society must inculcate its own ethics.

The task of instilling military ethics can be examined from at least three perspectives: (1) in terms of the difficulties involved in teaching military ethics; (2) from the point of view of the institutional forces that would have to be put in place in order to support ethical values and their application on a day-to-day basis; and (3) in terms of the enforcement mechanisms required in order to ensure that those who do not live up to the ethical code are not permitted to remain within the profession for long. By using these three perspectives as an organizational format, the problem of inculcating ethics within the military can be examined.

Teaching Ethics

In teaching military ethics, some basic assumptions about the nature of military ethics must be understood. The first is that the military is a profession and is not just another occupation in the larger society. The values it maintains are separate and different from those of the civilian sector. As a profession, it has at its core a sense of special calling that demands special obligations from its membership. Membership in a profession is defined precisely in terms of the members' willingness and ability to observe and bear the special obligations levied upon them. Those who are unwilling or unable to endure the special obligations and burdens must leave. It is never a question of the profession changing what it regards to be its ethical center simply to increase recruitment or sustain numerical strength. Thus, the first task in teaching military ethics is to spell out the specific obligations of the membership. There is, therefore, a need for a formal code of military ethics.

The first step in generating any code for any profession has always been to clarify and write it down. From the times of the ancient Egyptians when the imprimatur of the Pharoah was "so let it be written, so let it be done," until the present, those who would formulate codes of behavior have always found it necessary to inscribe them. In order to instill military ethics into the profession of arms, we must first be clear as to what they mean and what they require, and that means that a code of military ethics must be formalized. Without a code military ethics cannot be taught in any meaningful sense.

Once a code of military ethics has been established, the next task is to construct appropriate courses for teaching and understanding the code among members of the profession. Perhaps the simplest place to begin is at the service academies because they represent essentially schools for military professionals. As four-year colleges, they have adequate time to become involved in the design and testing of courses in military ethics. In any case, those courses which would apply at the academies could serve as models for wider use throughout the profession.

The service academies should institute at least four courses specifically designed not only to teach the profession's ethical code but also to train the soldier's faculties of ethical reasoning. While other parts of the curriculum might support the courses in ethics, the four proposed courses would be designed specifically to create within the curriculum an independent field of study dealing with military ethics. Ethics has too long been an appendage of or an adjunct to other fields of study.

One of the four courses would be a full year's course on the history of ethical thought. It would present a full range of developed theories about world philosophy and ethics, serving essentially as an introduction to philosophy, with emphasis on ethical questions raised or implied by the different theories. The course would provide every member in the curriculum with a common background and an informational base as to the kinds of questions that have plagued ethicists throughout the ages. Moreover, it would be a course in descriptive ethics, presenting a whole range of ethical theories and philosophies and making no real attempt to suggest that any one view is better than or preferable to another. This one-year course in the history of ethical thought would closely parallel courses found at the military academies now.

A second course would be a one-semester course of study in ethical reasoning. Here the emphasis moves away from descriptive ethics towards an analysis of ethics and the ethical reasoning that supports them. Training would be required in the mental processes of argument and logical reasoning. The assumption here is simple, that an officer or any member of the profession not only must be aware of the precepts of an ethical code, but must also know why these precepts ought to be observed, namely, why they bind as they do. Since a soldier can never avoid ethical judgments, which involve making choices among obligations, the soldier must be able to know the "why" of his action. It is impossible to instill military ethics or any ethics in anyone without some training in ethical reasoning. To be sure, ethical reasoning in itself does not assure that one will act ethically all the time. But as Derek Bok has pointed out, "formal education will rarely improve the character of a scoundrel. But many individuals who are disposed to act morally will often fail to do so because they are simply unaware of the ethical problems that lie hidden in the situations they confront."[1] One great shortcoming of the military profession is its members' lack of training in ethical reasoning. As a result, they are seldom able to recognize ethical dilemmas, much less reason their way through them. The argument along ethical lines is crucial to the inculcation of ethics in any profession.

A third course of study, also of one sememster's duration, would be a formal course in military ethics in which the precepts of the profession's ethical code would be taught as constituting the ethical center of the profession which all members would be required to observe and understand. The course would attempt to develop the code, explain why it is worthwhile in itself, and why it ought to be obeyed. The course would

Instilling Military Ethics 209

constitute the centerpiece of the program of ethical instruction. To be eligible for this course, students would be required to have taken the courses in the history of ethical thought and ethical reasoning, so that they would not come to the course merely to learn to obey as much as to learn and understand why the code has value. This course would provide the profession with a clear opportunity to state openly the ethical obligations it expects its members to observe.

The profession's ethical code is not to be offered as just one other ethical theory that the members can accept or reject. Those who take the course of instruction must be made to understand that the military code of ethics represents the ethical core of the profession of arms. Thus, the presentation of the code would take on a quality of normative indoctrination. Indeed, when taught at the staff schools and at the basic training level for enlisted soldiers, it would become a course in indoctrination. The point is that members of the military must be made to see through the use of reasoned argument and intellectual analysis that the ethics of the military profession is necessary to its proper functioning and to the proper behavior of the soldier. No one should be allowed to remain in the profession who cannot agree with the code or who cannot comprehend the reasons for it. As the ethical center of the profession, belief in the code and a willingness to abide by it would define membership in the special profession of arms. It is a contradiction to say that one ought to have within a profession a member who disagrees with its fundamental ethical precepts any more than one would tolerate within a monastery a monk who kept a harem: One cannot at one and the same time be a legitimate member of a profession and reject its ethical code.

This course in military ethics would provide intellectual exercise, a demonstration of the value of the code, and indoctrination. It would convince the soldier that this specific set of ethical precepts constitutes the center of his special profession and that, as ethics, they have value for him as a soldier. No code of ethics that does not affirm it is better than some other code or does not have special applicability to the professional task will have much attraction. Indeed, it cannot hope to be a code of ethics at all. If the military is not prepared to suggest to its members that those who do not believe in and observe the code must leave the profession, that its code is better than others, and that freedom of ethical choice does not include the ability to select contrary values, and if it is not prepared to enforce its position, it is not teaching ethics at all.

Finally, there should be a one-semester course in ethical dialectics in which members of the profession, using the seminar and Socratic approach, would be forced to solve ethical dilemmas on a case study basis. The emphasis would be upon the *process* of solution and the reasons for accepting certain solutions over others. The use of the infamous "school solution" is meaningless unless utilized to portray the reasoning that underlies the solution. The free interchange of ideas and challenges to the code would be tolerated and encouraged. The point of the instruction in ethical dialectics would be to sharpen the ethical reasoning skills of the soldier in a context where the examples used would be drawn as applications of the code under various circumstances. The problems, examples, and solutions should be related as much as possible to the kinds of ethical problems (combat and noncombat related) one is most likely to face as a soldier. Here, a heavy dose of realism buttressed by honesty would be needed. In short, it ought never to be forgotten that the course deals with the ethics of the soldier, and so ethical instruction should be appropriate to the ethical problems the soldier is likely to face.

Teaching these four courses in the sequence recommended here could easily be done at the service academies and at those military colleges from which many of the profession's officers are drawn. At the ROTC level, the profession's largest source of officers, one might expect that given the pressures of the college curriculum in nonmilitary subjects some reductions would have to be made. At the very minimum, ROTC cadets ought to be exposed to the course in military ethics and the course in ethical dialectics. Yet, somewhere in their college education most ROTC students are likely to be exposed to at least one course in introductory philosophy or a humanities course, or even a course in Western civilization which would acquaint them with competing ethical theories. Since ROTC programs cannot be as totally structured as those of the service academies or military colleges in terms of available time, the profession will have to make do with exposing ROTC officers to at least the courses in the ethical code and ethical dialectics.

If ethics is to be dealt with properly within the profession, ethics courses should be utilized at the staff schools and the war colleges. The Clement and Ayers study on the ability of the service staff schools to deal with the problem of ethics found that the students "lack conceptual understanding of professional ethics to address issues substantively, and that they also lack a common vernacular to communicate on the substantive issues."[2] Clement and Ayers also observe that even when courses in ethics are offered they are normally offered as electives. Within such courses, discussions of ethics usually degenerate into "war stories" in which superiors are blamed while "underlying issues are left unaddressed."[3] Merely exposing West Point cadets or ROTC officers to a few courses in military ethics at the beginning of their careers and then letting it go at that will not be sufficient for the profession to sustain its sense of ethical bearing. All of the service staff schools as well as the war colleges should be required to offer mandatory courses in ethics designed to keep an officer's ethical reasoning skills sharp and his ethical sense functioning. A course in ethical dialectics should be offered as a minimum requirement at these schools, with perhaps a shortened course on the ethical code as a prerequisite. It might be possible to combine the two with part of the course used to restate the problems of military ethics and the precepts of the code, and then combine it with a series of seminars in ethical dialectics in which the officers attempt to reach solutions.

Requiring courses in ethics (which is not the case now) would signal to the profession, as well as to those outside it, that military ethics is to be taken very seriously. At every level of professional instruction, each military curriculum to which officers and NCOs are exposed should include at least a course in the military ethical code.

An important additional step must be taken. Besides teaching ethics, there is the need to promulgate a code and to install mechanisms to support its observance. The attempt to instill ethics in the soldier is hardly a novel venture, and, in the United States, we have had some direct experience with the institutionalization of codes for the soldier. In 1953, largely as a reaction to the poor performance of American POWs in the Korean War, President Eisenhower ordered the creation of the now famous Code of Conduct. The Code of Conduct clearly and formally states the military profession's expectations as to how POWs ought to conduct themselves while in the custody of the enemy. Having designed the code at the insistence of the president, who himself was a former general, the military set about to teach it, and it did so very easily. It first formalized the code. It devised pocket-size cards on which the code was printed and required that all soldiers carry them; it developed propaganda posters that were displayed in barracks and other appropriate places throughout the military; it designed and issued appropriate plaques and documents enshrining the code; and it gave all soldiers at all levels of training several hours of basic instruction in the meaning of the code. Interestingly, the code was taught along with the reasons why it was binding. The soldier was not only taught how to behave, but also why he ought to behave in a certain manner. Besides being required to carry the card at all times, the soldier was required to memorize it. (The ability to recite a given precept upon request became a fixture of military inspections.) And the process worked.

Considering this experiment, there is no reason, then, why a code of ethics cannot be promulgated and printed in the form of cards, posters, and plaques to be displayed throughout the military establishment. There is also no reason why several hours of instruction concerning the code and its meaning cannot be accomplished at the basic training level, the NCO academy, and, more extensively, at the officer level. There is also no reason why the soldier should not be required to carry the code with him, to memorize it, and to be held responsible for observing it. If one can judge the effect of the Code of Conduct by the behavior of American prisoners taken in Vietnam, except for a very small number the behavior of American prisoners was exemplary in instances bordering on valor. The record was much better than that in Korean prison camps where our soldiers had not been exposed to a code of conduct prior to capture. There is at least some tentative evidence, then, that the establishment and promulgation of a code of ethics can work, that it can be accomplished very simply, and that it can produce beneficial results.

At risk of oversimplification, if one is going to articulate a military ethic and invest it with a sense of value in the eyes of the soldier, so that the soldier can internalize ethical values, then one must teach military ethics. Accordingly, there must be the formulation of a code, the construction of teaching mechanisms, and the emphasis placed not upon rote learning but upon teaching the soldier why the ethical precepts of a code ought to be observed. The level of sophistication of this effort will, of course, vary as one moves from the basic enlisted level through the NCO corps to the officer corps. But developing the ability to teach military ethics should not be that difficult for the military itself. The military did it before with the Code of Conduct, and there seems no good reason why it cannot do it again. But, of course, it must first be willing to do so, and that raises the question of the kind of institutional forces that must be dealt with so that the military bureaucracy will be able to teach military ethics.

Institutional Forces

How do we inculcate new values into the military profession and then ensure that the members, especially the officer corps, observe them? It must be remembered that the military bureaucracy is more of a corporate than an entrpreneurial entity, although the strains of the latter are clearly evident within it. It is more a profession than a business occupation with all that implies in terms of the characteristics and limits upon individual behavior that have already been addressed. Furthermore, military bureaucracies are, strictly speaking, institutions that are "value infused" as opposed to organizations that are mere instrumentalities for marshalling human energies to accomplish tasks. As institutions and not mere mechanisms, they have a "history" with a large investment in existing practices and values. Over the last twenty years, the American military profession and its bureaucracy have invested heavily in managerial and entrepreneurial values. Any proposed change in ethical values that runs contrary to the history of the institution will not easily gain support among the membership, especially among leaders whose very positions derive from support of the old values.

Finally, a distinction is in order between the adoption of new values on the one hand and latent and cognate values on the other. Latent values already enjoy the *public* support of the bureaucracy, even though they are not very effective in compelling individual behavior. Cognate values are those values that can be directly and obviously deduced or implied from existing values. The point is that there is likely to be greater institutional resistance to totally new values than to latent or cognate values. In the case of the cognate values, it may only be a matter of making already acceptable and supported values operative in terms of stimulating organizational and individual behavior.

So it would seem to be from the perspective of ethical values. Most of the values offered as ethical precepts for the military throughout this treatise are in fact not new to the military at all; they generally fall into the category of latent and cognate values. They already enjoy wide support within the profession, if only in terms of formal lip-service even while they do not compel much behavior. The task is made somewhat easier in that the values one is asking the profession to implement are essentially the values it should be implementing already as a profession. It will be somewhat easier to get the military to reassert its latent values than to get it to adopt and assimilate entirely new values. Given these few preliminary distinctions, what institutional forces may play a role in getting the military to adopt new ethical values?

One of the most important variables in the change process is the extent to which the profession can marshall the open, formal, and forceful support for new values from the elites positioned at the highest levels within the profession itself. For the military bureaucracy this means that the highest levels of command and staff must lend the new values the full force of their authority and prestige. Ayers and Clement's study of leadership models of organizational ethics shows that subordinates look crucially to their superiors for role-models, especially in ethical training and example.[4] A number of studies of civilian and military bureaucracies have found, as did the Army War College Study cited earlier, that the conduct of one's superiors is among the most important variables in convincing the membership of any group to assimilate new values, especially new ethical values. Indeed, a 1971 study of civilian managers found that the conduct of superiors ranked second only to formal ethical codes as the most influential factor in producing ethical behavior among subordinate executives. It revealed that poor ethical example ranked first as a factor in promoting unethical conduct among subordinates.[5]

If the profession's leadership fails to support a change of ethical values, the membership will not know what their superiors expect from them and what the system will ultimately reward despite its public pronouncements. The Army War College Study documents a gap between the values to which members of the profession often pay lip-service, especially those men at the top, and the actual standards by which people behave and are judged. This gap must be narrowed, if not closed altogether. The primary responsibility for doing this rests on the bureaucratic elites positioned at the top of the profession. In addition, the leaders' open support helps clarify changes in policy and values, thus removing much of the ambiguity as to what behavior is expected of subordinates and what values the profession is exected to uphold.

Those who occupy the highest levels of command and staff and are charged with initiating and overseeing the transmission of new ethical values should share them themselves and make their feelings public. Above all, through their actions they must publicly support the new ethics, for the profession's membership, especially the officer corps, will be closely observing their superiors' conduct, at least in the initial stages of the transition to new values. One of the major conclusions of the Army War College Study was as follows:

Every junior officer that we talked to was looking so strongly at their senior officers for standards that they could follow that it almost hurt. The number of times that they felt that they have been let down by looking for higher standards from senior officers and not finding them is innumerable.[6]

How is the elite of the profession to be converted to new values when *ab initio* the success of their careers has been predicated on internalizing and adhering to the very values now under attack, namely, managerialism, careerism, and entrepreneurialism? This is a major problem. Two solutions are possible, neither of which is ideal. First, outside pressures can be brought to bear from other bureaucracies upon which the military profession is dependent. Civilian authorities within the Department of Defense, to say nothing of congressional or presidential authorities, must issue unambiguous directives supporting the new values and stipulating penalties for any failure to comply. The clear advantage of a military bureaucracy is that it can respond rapidly to civilian direction.

The second solution is a more immediate way of ensuring elite support: remove those elites who do not support the new values and replace them with men who support the new ethical code. This practice of the "circulation of elites" is not a new one to the American military, especially when a promotion to general officer rank is predicated on being on the "right" side of a policy debate. It was also at the base of the Scharnhorst reforms of the German Army in 1813 in which some 78 percent of the officer corps was forcibly retired. This process of accelerated circulation of elites is very common in other industrialized societies and even has its counterpart in the American practice of "deep selecting" officers ahead of their contemporaries for high positions. Admiral Elmo Zumwaldt's promotion to chief of naval operations ahead of some forty of his seniors is a good example. This practice has the additional effect of reinforcing the notion that the profession is sincere about enforcing new ethical values. There is no clearer statement of such intent than the removal of those at the top who refuse to comply with the new code of ethics.

Another variable in the process of changing values in a military bureaucracy is the use of strong indoctrination programs at all levels which teach, explain, and reinforce the new values and detail the kind of behavior expected of the soldier. The profession itself must teach its ethics to its members. These programs must be particularly strong at the entrance level, for it is the young officers and soldiers who can be expected to carry the profession's ethics throughout their careers and eventually to internalize them so that they become part of their personal value system. This approach, although it may strike some as radical, is in fact not very different from that followed in conducting ethics programs at the military academies and colleges. The concept of creating a "missionary corps" of young officers fired by a defined set of ethics that could be used to "leaven" the profession, although somewhat misdirected at present, is based on the notion that the future belongs to the young.

Middle and upper tier officers could, of course, be required to participate in a series of seminars on the new ethics that could be taught at the appropriate staff and war colleges. The exposure of junior officers and soldiers at the entrance level to new values places limits on the actions of those within the profession who do not observe the new ethical precepts. Faced with a junior officer corps indoctrinated in a clear code of professional ethics, existing bureaucratic elites will not be able to violate this code without serious risk of protest or exposure. To be sure, the military must be prepared to codify its new ethics, a proposition requiring a move away from the ethos of the business corporation.

No policy of value change is likely to succeed without the strong support of informal groups within the profession. Peer support is needed at all rank levels for the new values and the actions they require. This support requires time to develop, but should be based on the common experience of the membership and its exposure to the code, courses of instruction in military ethics, and the realization that observing the new ethics is the best way to become a successful officer or soldier. Without peer support, especially at the middle-rank levels of the officer corps, few individuals can be expected to stand alone in support of any values. Those in the profession must be aware that by adopting the code of ethics and acting in a manner consistent with the code the soldier will be rewarded for his honor and integrity rather than penalized for it as is sometimes the case.

Another factor important to the adoption of new values has to do with the perceptions of the soldier as a member of the community. The officer must perceive a link between supporting the new ethics, behaving correctly, and his own self-interest defined in terms of his chances to advance normally within the profession and in terms of achieving other goals relevant to the military profession. The self-interest of the individual soldier is never exclusively defined in terms of careerism, entrepreneurialism, and managerial self-serving. The assumption is that a wide range of goals and rewards relevant to and supported by the profession satisfy the individual soldier. The military made an enormous mistake when it reduced the wants and motivations of the soldier to pay raises, promotions, and other "hard" benefits, thereby force fitting him into the econometric model. The military has required that self-interest be defined in econometric terms in order to apply models of analysis adopted from the business community. It is time that self-interest be redefined in terms of communal obligations and values. The soldier must perceive that by observing the new ethical precepts he will be recognized as a good soldier and as a practicing member of the profession entitled to all the rights, privileges, and honors due a loyal member.

It is, therefore, evident that the enforcement of new value codes must consistently demonstrate to the membership that the values and their concomitant behavior strengthen communal and integrative links with their peers and superiors. It is by rewarding and honoring those who cherish the ethical precepts of the profession and who act accordingly that the military can signal to its members that it is honest and sincere about selecting those officers from its ranks who are not only the most technically competent but also the most heavily imbued with the honor and spirit of the profession itself.

Relevant to the success of the new ethics is the soldier's perception that his experiences, both personal and anecdotal, demonstrate that the new values are important to professional development and to survival within the institution itself. Official support in terms of rewards and sanctions must be consistent with the new values. If they are not consistent, the new values will quickly become latent or cognate values, and their impact on the behavior of the profession will become slight.

In a democracy, bureaucracies respond most rapidly when pressured by outside agencies. The presidency and the Congress are good examples of such outside institutions. In addressing the problem of how to establish new values within the military, one important element may well be the extent to which outside institutions exert support and pressure. It may be unfair to expect total "in-house" reform, although the military must initiate the appropriate actions to acquire a new ethical code. It cannot do nothing, claiming it can act only in response to congressional or presidential direction. The initial requests and support for a new code of values will have to emanate from within the military, and external forces will help finally achieve adoption of the code. The quickest route to a new code of ethics may lie in the military drafting a code and asking that it be overtly supported by an act of Congress or presidential directive. As mentioned earlier, this was essentially what was done in 1953 with the Code of Conduct.

The most important variable involved in grafting new values onto the military profession is time. The transformation of an organization into an institution takes time. This transformation is, among other things, a function of the adjustments that the organization makes to its own experiences. Out of these experiences it constructs a history which in turn greatly affects the manner in which its members perceive the world. It is this history which must be overcome if the introduction of new ethical values for the military is to be successful. Since the American military profession has spent the last twenty years adapting to the managerial values and practices instituted in 1960, it is unrealistic to expect this process to be reversed overnight. Furthermore, a profession's history can be modified only when the experiences that result from the adoption and application of new values are such as to provide the experiential base for the development of a new history. The critical element in this development is time.

If we assume a normal career span of twenty years and use 1960 as a baseline as to when managerial approach began in the military, it is evident that the overwhelming majority of members of the profession, especially those at the highest ranks, have served all of their time in a military that knows no other approach. This is not to suggest that all military elites are mere managers. The commentaries contributed by high-ranking officers on the subject of ethics strikingly reveal how much some of these men have retained their traditional nonmanagerial values. Many of these individuals have managed to reach the top without relinquishing these values. Nonetheless, the military profession as a bureaucratic institution is thoroughly dominated by values contrary to those historically associated with professions and contrary to those stated in the code of ethics offered in this treatise. Any attempt to reform the military's value system will require a strategy that can deal with the institutional forces identified in this section. Such reform requires a recognition of the hard bureaucratic realities and vested interests that support and prosper under the present value system.

Enforcing Ethics

As described above, instilling military ethics requires that ethics be taught throughout the profession and that certain institutional changes in support of the new values be launched. A third element in the equation remains to be addressed: enforcement. Enforcement fundamentally involves the ability of the profession to reward those who observe the code and to punish those who violate it. At the same time, the military cannot spend all its time punishing transgressors, nor can it afford to develop enforcement mechanisms that are overly legalistic and that suspend the judgment of the membership, only to substitute for it a sterile legalism. Instead, the military must make observance of the code attractive not only by offering career incentives, but also by punishing those who disobey its precepts. The profession must involve the membership intimately in the process of judgment and levy punishments in its name. The concept is not unlike that of pre-medieval Germanic notion of law in which membership in the community implied that one was willing to support the law and to participate in its enforcement. When members of the community violate the law, by their own actions, they place themselves beyond the community; they become in the original sense of the term "outside the law" or, more commonly, "outlaws" from the community.[7] This same law must operate in enforcing the code of ethics. Superiors and peers are responsible for enforcement, and any expulsion that occurs is ordered in the name of the community and the values it espouses. One of the more appropriate mechanisms for accomplishing this task is the honor court.

The honor court is a mechanism designed to bring peer pressure to bear on the transgressor, short of the actual legalistic enforcement processes commonly associated with Western court systems. The honor court is not to be a substitute for the legal adjudication systems now in evidence. Rather, it assesses offenses to the honor and spirit of the community but does not invoke the military court system.

What happens to a soldier or officer whose behavior has discredited the profession but who has committed no *technical* violation of the civil or military law? The military profession at present has the choice of taking formal action against the soldier, for which there are scant legal grounds, or simply doing nothing. A procedure that offers an alternative between these two extremes is needed. The alternative should ensure that the profession's ethical code is observed within the spirit of the community and that the profession take appropriate action against the transgressor without necessarily drumming him out of the profession. One such alternative which seems to work fairly well in Western and other countries is the honor court.

The honor court has a long history. It can be found in the Germanic tribal systems, which were essentially warrior societies, in the eighth and ninth centuries. The honor court system, at least as it reflects the absence of legalisms and the use of communal peer judgments, is also found in the Mongol Empire of the fourteenth century. The brotherhood of medieval knights clearly reserved to themselves the right to judge the actions of their peers, even when such actions were not in technical violation of civil law. More recent examples include the Germans, whose Scharnhorst reforms established honor courts throughout the German military system which were used until World War II. The regimental system of the British Army has an equivalent system in which a board of regimental officers meet and make recommendations as to the actions of individual officers. A common punishment is expulsion from the regiment. The Japanese Army under the emperor also had an honor court. In the United States, the three major military academies, as well as the military colleges which tend to pattern themselves after the national academies, have some form of honor court. Thus, the use of honor courts within the military profession is not a new idea. Given this tradition, why not establish and use an honor court system throughout the profession as a mechanism for enforcing and ensuring that the profession's ethical code is observed?

In 1956, the Soviet Army created a system of honor courts called the Officer Comrade's Courts of Honor specifically designed as a social device to instill and enforce a sense of proper values and behavior for the Soviet officer. It is ironic that in a totalitarian society whose political regime extends to the control of all aspects of life that the Soviet regime should insist that the military itself be responsible for ensuring its own proper behavior and sense of honor. The Soviet honor court investigates offenses that discredit the rank and reputation of the military profession, that violate military honor, and that are inconsistent with the Soviet's view of civic morality. The honor court does not judge criminal offenses, and in no sense does it replace regulations or criminal codes, or serve as a mechanism for dealing with violations of discipline. As in other armies, there are adequate judicial guidelines and mechanisms for dealing with such problems. The honor court is convened to investigate an officer's conduct upon request of the commander under whose sanction the court has been established and only with the sanction of the offender's immediate superiors. In the Soviet system, honor courts are established at regimental and division levels.[8]

The manner in which these courts operate is a classic example of the way honor courts have historically operated, an example which could easily be followed in the American context. Soviet honor courts are not legal instruments of the state. Rather, they are instruments of social and peer pressure and are based on the proposition that the military ought to be the watchman of its own conduct. The courts are comprised of officers elected at the level at which the court has jurisdiction. Separate courts exist for junior officers up to the rank of captain, with separate courts for field grade officers and above. Hearings are conducted in public in order to facilitate peer pressure and to maximize the exposure of the offense to one's peers. Attendance at an honor court is regulated by rank; no one below the rank of the accused may attend. Thus, if a captain is being accused, no officer below that rank may attend. No permanent documentation is kept. The idea is to use peer pressure as an informal mechanism for bringing to bear the sense of the professional community in judging offenses against honor and ethics, conduct which, although perhaps not technically illegal, discredits the profession itself.

Courts of honor do not levy criminal penalties and have no ability to enforce any of the recommendations they make. Again it must be understood that honor courts are not substitutes for criminal or military courts or intended to deal with violations of law, regulations or discipline. The honor court is an institutional representation of the fact that not all violations of a code of military ethics are legal violations or violations of military regulations and discipline. Indeed, the distinction between codes of ethics and codes of law is fundamental to the operation of honor courts. Accordingly the court does not levy criminal penalties. What it does do is formally draft its findings and recommendations, make them public, and send them on to the next higher authority or the offender's commanding officer for action. In the Soviet honor court system, the court may make several recommendations. It may recommend that an officer found in violation be reprimanded or publicly sanctioned for the offense and that he be passed over for promotion or even reduced in rank. The court can recommend that a dishonorable officer be transferred to the reserves outside of active service or to another unit, or that he be separated from the profession entirely. These recommendations are made publicly by a board of officers and peers acting in judgment of the conduct of one of their brothers. They may be implemented or ignored by the offender's commanding officer, but the fact that they have been made, that one of the profession's own has been found wanting in the eyes of his brothers in terms of standards defining the community, cannot be ignored.

At the very least, the Soviet honor court system works to discourage dishonorable behavior through the threat of public disclosure, and so these courts represent an effective way of institutionalizing the core ethic of the military. A code of military ethics that is not enforced is likely to be honored more in the breach than in the observance. If there are no penalties for violating the code, it is likely to take a secondary position to narrow legalistic interpretations of a soldier's responsibilities. An informal court of honor would provide the profession with an easily accessible mechanism for interpreting and applying the code to members of the profession. The fact that the Soviets have used it for twenty-five years without serious abuse may or may not be a good recommendation for its adoption. The fact that the honor court has been used throughout Western history suggests that it should at least be explored and given serious consideration as a means of dealing with members who are unable or refuse to live up to the special obligations that constitute the ethical heart of the profession of arms.

Conclusions

In order to institutionalize a code of ethics, the military must reject any characterization of it as equivalent to a mere occupation or an "employer of last resort." If the military is merely another job, then there is no need to levy upon it the burden of a special calling. If the military is able to define itself as a profession it must next be able to agree on the values that comprise its core ethic by formulating a code of ethics for its membership. It is nonsense to affirm that a subsocial group is a profession with a required sense of special obligation and then plead ignorance as to the ability to formulate a code of values and behavior that rests at the center of that sense of obligation. Until a profession can clarify its own role, it will remain uncertain of its values, and until it can develop a code of ethics for its membership, it will be difficult to institutionalize a sense of its own ethical worth.

But, if the military is to institutionalize a code of ethics, it must be able to establish instructional courses of study in ethics at various levels of sophistication. The soldier cannot be held responsible for what he ought to do unless he is told what he ought to do. Moreover, if the military profession is required to be somewhat separate from the larger society, the novice to the profession must be informed of the special obligations required of him as a member of the profession. There is, then, the initial need to teach ethics, as well as to continually reinforce this ethical sense through the use of appropriate rituals and symbols. The promulgation of a code of military ethics along the lines of the familiar Code of Conduct might be further buttressed by requiring every soldier to reaffirm his oath at least once a year in some appropriate ceremony, perhaps on the birthday of his respective service. Symbolism and ritual can be powerful mechanisms for reinforcing and sustaining values, especially in the military where such rituals can be directly associated with the most potent of anthropological forces, life and death.

If the military is to attempt to institutionalize a code of ethics, it must be prepared to reform the existing bureaucratic apparatus so as to weaken those institutional supports capable of resisting the new values. It is necessary to restructure the organizational apparatus of the military so as to develop institutional mechanisms for promulgating, sustaining, and enforcing a code of military ethics. Without such reforms in the bureaucracy, new values will not be adopted.

Finally, adequate enforcement mechanisms are crucial. In the enforcement of a code of military ethics, it must be understood that a code of ethics is not the equivalent of a body of law and that its enforcement mechanisms are not the equivalent of courts of law. Mechanisms for enforcing ethical codes within the profession should consist of formalized procedures for bringing to bear the consensus of the profession as to the obligations the membership must observe. To confuse law with ethical codes is a mistake. The honor court is a mechanism for enforcing a code of ethics and not a body of law. To confuse this point will inevitably lead to the degeneration of the honor court into one more forum for the spouting of sterile legalisms. There must, therefore, be a sense of communal trust among the membership of the profession to enforce the code with judgment and justice. If the military cannot trust its own membership to judge its own, to discipline its members, if it cannot trust its officers to behave ethically in pointing up and judging the suspect behavior of others, no amount of legal guarantees or lawyers will be able to engender that sense of trust for us. In this sense the military is akin to the Germanic tribe, akin to the *communitatus* of the medieval knight, and akin to the communal brotherhood of the monastery.

The task of ethical reform within the military is not an impossibility. It has already been carried out in other armies. If the German, the Japanese, and the British military establishments can establish honor courts and if we can establish them at our military academies, we ought to be able to establish them and make them work throughout the profession. Of course, the ability to carry out the task of formulating, instilling, and enforcing an ethical code must rest with the military itself.

The direct responsibility for these actions rests most clearly with the highest ranking leaders of the profession. Unless they are willing to provide the institutional support and, above all, the moral example to move the profession in the direction of constructing and enforcing a code of ethics, very little is likely to be achieved. History records no examples of ethical reforms that were self-generating or self-sustaining. The problems of the profession were made by men and they can be undone by men. The ability to set standards which ennoble men in their striving to attain them is perhaps one of the greatest qualities of the human character. It is this quality that must be brought to bear upon the problem of ethics by the top elites of the military profession.

In a democratic society based on capitalistic values, if the military does not take steps to insure its own sense of honor and ethics, the larger society cannot bear this burden for the profession. The importance of honor can never be underestimated. As one Canadian officer put it, "all the soldier can bargain with is his honor and the honor of the soldier, like the virtue of a maiden, once taken cannot be restored."[9] It is the responsibility of the profession itself to set high ethical standards by which men may live, for it is only in setting high standards that the sacrifice which we require of the soldier can ever be justified.

Notes

1. Derek C. Bok, "Can Ethics Be Taught," *Change* (October 1976): 28.
2. Stephen D. Clement and Donna B. Ayers, *A Matrix of Organizational Leadership Dimensions* (Fort Benjamin Harrison, Ind.: U.S. Army Administration Center, 1976), p. 21.
3. Ibid.

4. Ibid.
5. Steven N. Brenner and Earl A. Molander, "Is the Ethics of Business Changing," *Harvard Business Review* (January-February 1977): 64.
6. *U.S. Army Study on Military Professionalism* (Carlisle Barracks, Penn.: U.S. Army War College, 1970), p. 30.
7. Fritz Kern, *Kingship and Law in the Middle Ages* (Oxford: Basil Blackwell, 1968), Part II.
8. "Soviet Army Discipline," *Military Review* (November 1975): 46.
9. Major R. I. Aitken, "The Ethical Aspects of Professionalism," *Canadian Forces Staff School* (unpublished paper, April 1979), p. 17.

9 FINAL THOUGHTS

Throughout the long history of warfare, the only constant that emerges is the nature of man. Technology, weaponry, strategy, and tactics may change, but the man who implements them has not. The men who stood at Thermopylae in ancient Greece were no different than the soldiers who stood at Waterloo, the Somme, Bastogne, Heartbreak Ridge, and Khe Sanh. Their fears, hopes, and expectations were the same. Their expectations of themselves, their peers, and their officers were the same. For good or ill, man's nature has not changed in the four million years of his life on this planet—nor, indeed, has he foregone its most obvious manifestation, war.

The technology of the modern age raises questions about its impact on the conduct of war, just as the arrival of new weaponry provoked the same questions in earlier periods. That technology will have an effect on the modern battlefield, perhaps in the form of increased lethality, is beyond question. That it will forestall war or remove the requirements that men in battle groups need to sustain themselves in the face of the horror of collective destruction is nonsense. In the equation of combat performance, technology will count for little if the men behind the weapons do not demonstrate the same qualities of courage, determination, skill, and composure in the heat of battle that soldiers armed with more primitive weapons demonstrated in earlier times. The long prophesied "push-button battlefield" has yet to arrive, and technology must still be placed in the service of brave men to be effective. It is not, nor has it ever been, a substitute for bravery and courage.

Yet, as we look to the battlefields of the future, the rate at which human lives will be consumed by the macabre dance of death will surely increase even in a conventional environment. The lethality of modern weaponry will increase the degree of battle stress that combat troops will be expected to bear, and the amount of time men will be able to sustain battle stress before becoming numbed by shock will shrink considerably. At the same time, the number of battle casualties due to psychological injury will rise dramatically. Yet, the battle must be joined. Under these conditions, the requirements for high levels of unit cohesion will increase, as will the need for those elements that sustain cohesion. Without unit cohesion as a shield between the soldier and his hostile environment, whole armies will be unable to function on the battlefield, regardless of the level of technology they can place in the hands of their soldiers.

Crucial to the ability to bond units together under stress is the need for ethics. There must be clear evidence among men in battle groups that their peers and superiors are living up to their obligations if the soldier is to live up to his. In an environment filled with common horror, a belief in the values of the profession becomes critical to psychological survival. It is this belief that gives meaning to life. Without a sense of ethics at the core of the profession to levy obligations which members believe will be observed, the combat units of the future will find it difficult to withstand the rigors of warfare.

War is a very human act, an act that has been repeated in all human societies since the very beginning of man's history. Perhaps the tendency toward warfare and conflict is sown into our very genes; certainly no society has been able to make do without the presence of the soldier. While the work to eradicate war should proceed, it seems unlikely to succeed. If war is a constant in human affairs, it can have limits only if the men who fight it have a sense of meaning and limit. A sense of military ethics can give the profession of arms some import beyond the ability to destroy, and only a sense of ethics can limit the human destruction by clearly establishing a proper concert between means and ends. A soldier without ethics, values, and beliefs with which he can live in a moral sense will himself be destroyed by the horrors of the battlefield. Without ethics the humanist quality so necessary to moral man will die. Only ethics can place the destruction of warfare in perspective and prohibit men from using violence beyond reason. Without ethics the horror of human combat becomes even greater.

As long as men remain human, there will be a need for military ethics to sustain that humanity, to give meaning to actions that otherwise would be regarded as terrible, and to place limits upon the destructive abilities of the soldier. This is especially the case for the military forces of democratic societies whose *raison d'etre* is the pursuit of ideals and values based in human worth, ideals and values that separate them from the naked power upon which nondemocratic regimes are based. For the military profession to fail to develop and teach a code of ethics is to ignore a crucial obligation to its membership, to make them good in the exercise of a profession that often directly confronts the face of absolute evil. As with other obligations, there is no escape from this one either. We dare not do battle without a sense of ethics at the center of the profession, for if we do the chances are great that we will destroy that which we seek to preserve by our efforts, a sense of humanity among men.

There is yet another danger. In an age of increased complexity, specialization, and bureaucracy, the propensity to limit man's judgment or to confine it to excessively narrow tunnels of expertise or, in some instances, even to do away with it altogether by subordinating it to prearranged rules and regulations raises the danger that the soldier will lose sight of the purpose of his profession and the reasons for his sacrifice. There is the danger that, lost in a forest of complex systems, he will accept his escape from freedom and willingly allow his judgment to be suspended and replaced by the dictates of his superiors or the organization. Under these circumstances, the soldier risks becoming an expert in the application of violence who plies his trade in the service of the state with little or no regard for the ethical concerns that might be involved. As the exercise of ethical judgment becomes more difficult, as life becomes more organized, the temptation to abandon ethical judgment altogether may prove overwhelming. If so, the military profession will bear witness to its own death as a profession, and the commission of ethical horrors will not be far behind.

The profession of arms is by its nature involved with special ethical responsibilities. Any profession that deals with the lives and deaths of its fellow citizens on such a large scale carries with it heavy burdens, more so in an age where whole societies may be executed by a single stroke. Only a sense of ethics, of right and wrong, of limit and perspective, and of special service can forestall the degeneration of a noble profession into a senseless purveyor of violence. It will avail a free society little if in the effort to protect its freedoms it allows the creation of a military force whose actions violate a clear sense of ethics. Over time, such a force will become increasingly out of step with the society it serves, becoming one more self-interested pressure group within it and inevitably a threat to it, if not by overt action, then by poor example. There can be no question of an unethical military profession serving an ethical society. One will surely corrupt the other.

We dare not continue down the road upon which we set out some two decades ago. We cannot allow the continued erosion of the military away from its traditional values and norms as a profession, only to see them replaced by the corrosive values of managerialism and careerism. We cannot abide the transformation of the military from a profession towards an occupation, for to continue to do so is to make one of the most morally difficult acts, the taking of human life, merely an ordinary job. Furthermore, it turns the sacrifice of those who, in the service of something noble, lived up to their unlimited liability and gave their lives into a mockery of economics. It is unacceptable that we should count the sacrifice of brave men in the service of ideals as merely one more "expenditure of resources" or as merely the "cost of doing business." But that is all we are left with if the way of the soldier is allowed to turn from a profession into a mere occupation.

If we are to stop the erosion of our ideals and rebuild the profession, if we are to steel the psyche of the soldier against the horrors of battle, if we are to give meaning to the sacrifice of those who have gone before, and if we are to expect those now in service to follow their example even unto death, we must reconstitute the brotherhood of arms along lines of ethics. Without a code of ethics, the task of the soldier degenerates into senseless violence.

We must be keepers of our own flame for it rests with us—the officers and men of the profession—to bear the burdens of sending our fellow citizens to their deaths and trying to live with their ghosts. The burden will become intolerable unless we are certain that what we do is right and unless the guideposts for such judgments reside in a code of ethics. We dare not face the future, as we have faced the recent past, without a code to sustain us. The world is already too much with us as external values eat at the core of our profession. Without a code of ethics we may still fight, but we will fight as hollow men uncertain of our country, uncertain of our profession, but, most of all, uncertain of ourselves and our humanity in a world already grown too inhuman.

BIBLIOGRAPHY

Aitken, R. I. "The Canadian Officer Corps, the Ethical Aspects of Professionalism." *Canadian Forces Staff School* (unpublished, April 1979).

Bachman, Gerald G., and John D. Blair. "Citizen Force or Career Force?" *Armed Forces and Society* 2, No. 1 (November 1975):81-97.

Berger, Peter. *Pyramids of Sacrifice: Political Ethics and Social Change.* New York: Anchor Books, 1976, p. 249.

Berry, F. Clifton, Jr. "A General Tells Why the Army Is Its Own Worst Enemy." *Armed Forces Journal International* (July 1977):22-23.

Beveridge, J. R. "In Defense of the Regimental System." *From Our Readers* (August 1958), pp. 45-47.

Blumenson, Martin. "Some Thoughts on Professionalism." *Military Review* (September 1964):12-16.

Bok, Derek C. "Can Ethics Be Taught?" *Change* (October 1976):26-30.

Bradford, Zeb B., Jr. "Duty, Honor, Country Vs. Moral Conviction." *Army* (September 1968):42-44.

Brenner, Steven N., and Earl A. Molander. "Is the Ethics of Business Changing?" *Harvard Business Review* (January-February 1977):57-71.

Brown, Frederic J. "The Army and Society." *Military Review* (March 1972):3-17.

Brown, Reginald J. "The Meaning of Professionalism." *American Behavioral Scientist* 19, No. 5 (May-June 1976):511-522.

Bunting, Josiah. "The Conscience of a Soldier." *World View* 16, No. 12 (December 1973):6-11.

Caiden, Gerald E., and Naomi J. Caiden. "Administrative Corruption." *Public Administration Review* 37 (May-June 1977):306-307.

Carroll, Archie B. "Linking Business Ethics to Behavior in Organizations." *S.A.M. Advanced Management Journal* (Summer 1978):27-29.

———. "Managerial Ethics: A Post-Watergate View." *Business Horizons* (April 1975):75-80.

Carroll, Robert C. "Ethics of the Military Profession." *Air University Review* (November-December 1974):39-43.

Clark, Wesley K. "The Elusive Concept of Honor." *Armor* (September 1971):22-25.

Clement, Stephen D., and Donna B. Ayers. *A Leadership Model for Organizational Ethics*. Fort Benjamin Harrison, Ind.: U.S. Army Administration Center, 1976, p. 21.

Clotfelter, James, and B. Guy Peters. "Profession and Society: Young Military Officers Look Outward." *Journal of Political and Military Sociology* 4 (Spring 1976):39-51.

Cogar, Lawson G. "The Officer's Ethical Obligation." *Marine Corps Gazette* 61 (February 1977):56-57.

Combs, Cecil E. "On the Profession of Arms." *Air University Review* 15, No. 4 (May-June 1964):2-9.

Deagle, Edwin A., Jr. "Contemporary Professionalism and Future Military Leadership." *Annals of the American Academy of Political and Social Science* (March 1973):162-172.

Deininger, David G. "The Career Officer as Existential Hero." *U.S. Naval Institute Proceedings* (November 1970):18-22.

Drisko, Melville A., Jr. "An Analysis of Professional Military Ethics: Their Importance, Development and Inculcation." U.S. Army War College Study Project. Carlisle Barracks, Penn., June 19, 1977.

Duncan, J. W. "Trust and Confidence Revisited." *Marine Corps Gazette* (November 1976):60-61.

Dunwell, Ronald P. "Erosion of an Ethic." *U.S. Naval Institute Proceedings* (March 1977):57-62.

Dyck, Arthur J. "Ethical Bases of the Military Profession." *Parameters* 10, No. 1 (March 1980):39-46.

Ebel, Wilfred L. "The Amnesty Issue: A Historical Perspective." *Parameters* 4, No. 1 (1974):67-78.

Eckhardt, G. S. "Study on Military Professionalism." U.S. Army War College, Carlisle Barracks, Penn., June 30, 1970.

Eggenberger, John. "Toward a General Model of Military Leadership for the Canadian Forces." Departmental Manuscript, Royal Military College of Canada, Unpublished, May 1979.

Ellis, Joseph, and Robert Moore. *School for Soldiers: West Point and the Profession of Arms*. New York: Oxford University Press, 1974.

Feldman, Fred. *Introductory Ethics*. Englewood Cliffs, N.J.: Prentice-Hall, Inc., 1978.

Flammer, Philip M. "Conflicting Loyalties and the American Military Ethic." *American Behavioral Scientist* 19, No. 5 (May-June 1976):589-603.
Frankena, William K., *Ethics*. Englewood Cliffs, N.J.: Prentice-Hall, Inc., 1973.
Futernick, Allan J. "Avoiding an Ethical Armageddon." *Military Review* (1979):17-23.
Gabriel, Richard. "About Face on the Draft." *America* (February 1980):95-97.
———. "Acquiring New Values in a Military Bureaucracy: A Preliminary Paradigm." *Journal of Political and Military Sociology*. (Spring 1979):89-101.
———. "Ethics and Military Community." The Hastings Center Report, August 1980.
———. "Legitimate Avenues of Military Protest in a Democratic Society." *Journal of Professional Military Ethics* (April 1980):2-9.
———. "Military Structures and Combat Cohesion: Lessons for the Canadian Regimental System." Royal Canadian Military Institute Yearbook, Winter 1979.
———. "Modernism and Pre-Modernism: The Need to Rethink the Basis of Military Organizational Forms." In Franklin Margiotta and Brown, *Military Organizations*. Colorado: Westview Press, Fall 1980.
———. Obedience and Resistance in the Political Thought of Thomas Aquinas: A View from the Bottom." *Delta Epsilon Sigma Bulletin*, Winter 1975.
———. "The All-Volunteer Force and American Values." *U.S. Naval Institute Proceedings* (September 1979), p. 22.
———. "To Serve with Honor: An Officer's View of the Ethics of Military Service." *Army Magazine* (May 1980): 8-12.
———. "Turning Away from Manageralism: The Environment of Military Leadership." *Military Review* (July 1980). pp. 55-68.
———. "What the Army Learned from Business." *New York Times*, April 15, 1979.
———, and Paul L. Savage. *Crisis in Command: Mismanagement in the Army*. New York: Hill and Wang, 1978.
———, and Paul L. Savage. *Managers and Gladiators: Directions of Change in the U.S. Army*. Boston: Hawkes Press, 1979.
Galligan, Francis B. *Military Professionalism and Ethics*. Newport, R.I.: Naval War College, Center for Advanced Research, June 1979.
Gard, Robert G., Jr. "The Military and American Society." *Foreign Affairs* (July 1971):668-710.
Gernert, William E., III. "On Fostering Integrity." *Air University Review* 27, No. 6 (September-October 1976):62-67.
Ginsburgh, Robert N. "The Challenge to Military Professionalism." *Foreign Affairs* 42, No. 2 (January 1964):255-268.
———. "Military Ethics in a Changing World." *Air University Review* 27, No. 2 (January-February 1976):2-10.

Goodpaster, Andrew J. "Concept for Furthering Cadet Moral Development." *Authoritative Guide*, July 15, 1978.
———. "General Goodpaster's Credo." *Paramenters* 5, No. 1 (1975):65.
———. "Moral Choices: Ethics and Values in the 80's." Speech made at West Point before the Association of American Colleges, Washington, D.C., February 4, 1979.
Guelzo, Carl M. "Chore or Challenge: A Professional Ethic for the Nuclear Age." *U.S. Naval Institute Proceedings* (May 1964):26-34.
Hackett, Sir John. *The Profession of Arms*. London: Times Publishing Co., 1962, p. 63.
Hart, T. S. "Determination in Battle." A presentation made to Drac Conference, November 29, 1978.
Hauser, William L. "Armies and Societies: Three Case Studies." *Military Review* 52 (July 1972):4-5
Hayes, Samuel H. "To Thine Own Self Be True." *Army* (July 1967):29-30.
———, and William N. Thomas. *Taking Command*. Harrisburg, Penn.: Stackpole Books, 1967, p. 51.
Huntington, Samuel. *The Soldier and the State*. Cambridge, Mass.: Harvard University Press, 1967.
Jesse, W. L. "The New Military Professional." *U.S. Naval Institute Proceedings* 101 (January 1975):35-31.
Johnson, Kermit D. "Ethical Issues of Military Leadership." *Parameters* 4, No. 2 (1974):35-39.
Kant, Immanuel. *Foundations of the Metaphysic of Morals*. New York: Liberal Arts Press, 1959.
Katzenbach, Edward L. "The Demotion of Professionalism at the War Colleges." *U.S. Naval Institute Proceedings* 91, No. 3 (March 1965): 34-41.
Keegan, John. *The Face of Battle*. New York: Vintage Press, 1976.
Kerwin, Walter. "Values of Today's Army." *Soldier* (September 1978):4.
Kindred, Jon D. "In God We Trust, Officers 'Post Bona'?" *Army* (December 1975):10-11.
Kinnard, Douglas. *The War Managers*. Hanover, N.H.: University Press of New England, 1977.
Leadership, unpublished pamphlet by the Canadian Forces Training Material Production Center. Canada, 1978.
Lloyd, Alan. *War in the Trenches*. New York: David McKay, 1977.
Loomis, Dan G. "The Regimental System." A Mobile Command Letter, special supplement. Unpublished.
Lynn, William M. "The Military Profession: What Is It?" *Army* (September 1971):23-27.

MacIsaac, David. "Where There's Pain There's Hope: Military Professionalism in the Dock." *Air University Review* 24, No. 6 (September-October 1973):93-102.

Mack, William P. "The Need for Dissent." *The New York Times Magazine*, January 12, 1976, pp. 20-26.

Margiotta, Franklin D. "A Military Elite in Transition." *Armed Forces and Society* 2, No. 2 (February 1976):111-184.

Meade, Henry J. "Commitment to Integrity." *Air University Review* (March-April 1977):86-90.

Moellering, John H. "The Army Turns Inward." *Military Review* (July 1973):68-83.

Moskos, Charles C., Jr. "From Institution to Occupation: Trends in Military Organization." *Armed Forces and Society* 4, No. 2 (November 1977):41-50.

———. "How to Save the All-Volunteer Force." *The Public Interest* 60 (Fall 1980):21-29.

———. "Serving in the Ranks: Citizenship and the All-Volunteer Force." Paper prepared for the Hoover-Rochester Conference on the AVF, Stamford, Connecticut, December 13-16, 1979. Unpublished.

Mussman, Arthur C. "The Unit Commander and the Bureaucracy." *Air University Review* 24, No. 6 (September-October 1973):83-88.

"On Ways of Thinking." *Manas* 32, No. 49 (December 5, 1979):1-2.

Palumbo, Dennis J., and Richard A. Styskal. "Professionalism and the Receptivity to Change" *American Journal of Political Science* 18, No. 2 (May 1974):387-388.

Paskins, Barrie, and Michael Dockrill. *The Ethics of War*. University of Minnesota, 1979.

"The Ragged Ranks of the U.S. Military." *Boston Globe*, September 9, 1980, p. 2.

Rosser, Richard F. "A 20th Century Military Force." *Military Review* (March 1974):42-54.

———. "Civil-Military Relations in the 1980's." *Military Review* (March 1974):18-31.

Russett, Bruce M. "Political Perspectives of U.S. Military and Business Elites." *Armed Forces and Society* 1, No. 1 (November 1974):79-101.

Sarkesian, Sam C. "An Empirical Reassessment of Military Professionalism." *Military Review* (August 1977):3-19.

———. "Moral and Ethical Foundations of Military Professionalism." Prepared for the Southeast Regional Conference, Interuniversity Seminar on the Armed Forces and Society—Air University, Maxwell Air Force Base, Alabama, June 3-5, 1979.

———, and Thomas M. Gannon. "Professionalism." *American Behavioral Scientist* 19, No. 15 (May-June 1976):495-509.

Savage, Paul L. "Historical Military Models and the American Army: Reform, Professionalism, Ethics and Conflict with Civil Society." Prepared for *The Citizen Soldier in Today's World*. St. Michael's College, Winooski, Vt., October 5-6, 1979.

Shoup, David M. "The New American Militarism." *Atlantic Monthly* (1969): 51-56.

Smith, Monroe T. "Reporting Inaccuracies—A Rose by Another Name." *Air University Review* 25, No. 2 (January-February 1974):83-88.

Sorley, Lewis. "Competence as an Ethical Imperative: Issues of Professionalism," Paper presented at the Southeast Regional Conference at Maxwell Air Force Base, Alabama, June 3-5, 1979.

———. "Duty, Honor, Country: Practice and Precept." *American Behavioral Scientist* 19, No. 5 (May-June 1976):627-645.

"Soviet Army Discipline." *Military Review* (November 1975), unclassified.

Stackhouse, Max L. "Military Professionalization and Values." *Military Review* (November 1973):3-8.

Stockdale, James Bond. "Taking Stock." *U.S. Naval War College Review* (February 1979):1-2.

———. "The World of Epictetus: Reflections on Survival and Leadership." *Atlantic Monthly* (February 1978).

Stouffer, Samuel. *The American Soldier*. Princeton, N.J.: Princeton University Press, 1976.

Summers, Harry G. "The Lion Heads." *Military Review* (July 1972):107-109.

Taylor, Maxwell D. "A Professional Ethic." *Army* (May 1978):18-21.

Taylor, William J. "Military Professionals in Changing Times." *Public Administration Review* (November-December 1977):633-640.

Toffler, Alvin. *Future Shock*. New York: Bantam Books, 1977.

"The Officer." Unpublished pamphlet by the Canadian Forces Training Material Production Center, Winnipeg, Canada, 1978.

Toner, James H. "Sisyphus as a Soldier: Ethics, Exigencies, and the American Military." *Parameters* 7, No. 4 (1977):2-12.

———. "The Military Ethic: On the Virtue of an Anachronism." *Military Review* (December 1974):9-18.

Turkelson, Donald R. "The Officer as a Model of Ethical Conduct." *Military Review* (January 1975), pp. 20-31.

Ulmer, Walter F. "Diluted Professionalism." *Military Review* (December 1963):3-11.

Wakin, Malham M. "The Ethics of Leadership." *American Behavioral Scientist* 19, No. 5 (May-June 1976):567-589.

Webb, Ernest L. "NCO Corps. Is it Really Back on Top?" *Army* (July 1978):27-29.

———. "When Ethic Codes Clash: Absolute Vs. Situational." *Army* (March 1978):31-33.

Weigley, Russell F. "A Historian Looks at the Army." *Military Review* (February 1972):25-35.
Wenker, Kenneth H. "Professional Military Ethics: An Attempt at Definition." *Air Force Journal of Professional Military Ethics* 1 (April 1980):23-28.
Winters, Francis X. "Ethical Considerations and National Security Policy." *Parameters* 5, No. 1 (1975):17-25.

INDEX

Absolute ethics, 45-46
Academies, service, 5; scandals, 5; cheating, 5; honor code, 5; coverups, 5
Actual duties, 43-44
Alexander, Clifford, 69
All-Volunteer Force (AVF), 4, 5, 59, 68-70; desertion in, 69-70; discharges from, 70; quality of recruits, 69; social composition, 6, 96, 204
American cultural values, 15-18; ethical egoism, 17; Madisonian politics, 16; Smithian economics, 15; Social Darwinism, 16
Appeals to higher commands, 193-94
Army Order 100, 41
Army Readiness Study, 1980, 69
Army War College Study, 7, 71, 74, 75, 77, 175, 214
Art, 38
Assassinations, 4, 59, 64
Attrition, 70
Automated battlefield, 61

Bataan Death March, 192
Battlefield virtues, 170
Beard Report, 69
Bearing, 171
Body count, 4, 183
Bok, Derek, 20, 208
Brown, Reginald, 113
Bureaucratic ethics, 39
Bushido, code of, 200-201
Bunting, Josiah, 180-181

Cadet honor code, 120
Calley, William, 126
Careerism, 72, 98, 136-138, 183
Categorical imperative, 48-49
Centralized promotions, 13
Challenges to military professionalism, 94-107; bureaucracy, 99-100; competing ethics, 101-7; managerialism, 97-99; occupationalism, 94-97; specialization, 100-101
Character, 150-53, 120
Christian pacifist, 123
Civilianization, 11
Clause of unlimited liability, 61, 87, 144, 160, 181, 229
Code of conduct, 71, 211, 212, 218, 223

Code of ethics, 138-47; advantages of, 119-22; formal code, 140; objections to, 122-31; service academies, 83, 131-35
Codes vs. laws, 126
Cognate values, 213
Cohesion, 11, 59, 66, 69, 81, 227
Combat refusals, 4, 59, 64-65
Command, nature of, 141
Commitments, 104-7; external, 104; internal, 105-7
Communitatus, 224
Communal obligations, 217
Compassion, virtue of, 166
Competing obligations, 32
Confidence, virtue of, 169
Convergence, 109-10
Corporate values, 85, 107
Courage, virtue of, 170
Court martial, 199
Creative tension, 125
Crisis in Command, 98
Crisis of adaptation, 4
Crisis of confidence, 4
Crisis of conscience, 4
Crito, the, 26
CYA (Cover Your Ass) Syndrome, 162

Darwinian social ethics, 19
Decision of character, 171
Dedication, virtue of, 168
Defense Act of 1958, 131
DEROS system, 59, 66-68
Descriptive ethics, 37, 133, 208
Desertions, 4, 59, 64
Dien Bien Phu, 111
Dignity, virtue of, 171-72
Dissent, 162-63
Drisko, Melville, 75-77
"Drive for success," 12

Drug use, 2, 64, 69
Duty, nature of, 155-56, 197; perversion of, 197-98

Eichmannism, 24, 148
Eisenhower, Dwight D., 211
Elites, conversion of, 215; removal of, 216
Employer of last resort, 222
Ethical: action, 31; breakdown, 97; challenges, 11-13; choice, 28; commitment, 53; dialetics, 210; dissent, 183-85; egoism, 17-24, 101; handicaps, 14; imperatives, 49, 82, 122, 146; marketplace, 137, 147; precepts, 24; reasoning, 208; relativism, 11; renaissance, 8; responsibility, 40-48, 129; training, 145
Ethics, critical questions of, 9; descriptive, 19; prescriptive, 19
Ethics, definition of, 29; functions of, 25; need for, 56
Ethics of absolute ends, 46-48
Ethics of duty, 8, 151
Ethics of self-interest, 57
Ethics of service, 57
Ethics of virtue, 8, 157, 185
Existentialism, 101-7

Face of battle, 81
False reporting, 4
Fealty, 158-59
Fides, 73, 196
Flammer, Philip, 72
FM 27-10, 180
Freedom, 30
Free enterprise economics, 15
Free fire zone, 183

Gard, Robert, 89
German Army, 81, 111, 168, 177

Germanic tribes, 220-24
German law, 219
Gesinnungsethik, 46
Goethe, Johann, 153
Graf Spee, 200

Hackett, Sir John, 87
Hackworth, David, 184
Harvard Business School, 10
Hayes, Samuel, 159
Heartbreak Ridge, 226
Hochverrat, 196
Homo mensura, 47
Honesty, virtue of, 159
Honor, 157
Honor code, 131
Honor court, 219-23
Humanism, 165-66
Humanist, 85

Indoctrination programs, 215-16
Institutional forces in AVF, 11
Integrity, 154-55; ethical, 141; virtue of, 142
Intellectual curiosity, virtue of, 164
Introspection, 168-69
Invisible guiding hand, 15

Janowitz, Morris, 81
Japanese Army, 220
Johnson, Kermit, 212
Joint Chiefs of Staff, 176
Judgment, virtue of, 38, 154

Kant, Immanuel, 48-49
Keagan, John, 81
Kerwin, Walter, 91, 92, 94
Khe Sanh, 226
Kinnard, Douglas, 184
Korean War, 211
Koster, Samuel, 5, 12, 158, 192

Landesverrat, 196
Latent values, 213
Law of anticipated reactions, 189
Law of Land Warfare, 180
Liberalization campaign, 10
Lidice, 152
Loyalty: syndrome, 12; virtue of, 72, 158
Lynn, William, 73, 120

MacArthur, Douglas, 196
Mack, William, 74-75
McNamara, Robert, 6, 11, 61, 97
Madisonian politics, 16-21
Malum in se, 24, 26, 32, 46, 115
Malone, Mike, 154
Managerial tendencies, 6; values, 98
Marshal, SLA, 174
Marshall, George, 60, 156, 196
Mechanistic values, 107
Medieval knights, 172
Medina, Ernest, 192
Military, differences, 86; level of responsibility, 86-88; monopoly of practice, 87; personal liability, 87; scope of service, 86-87
Military: primary function of, 19; virtues, 150-51
Military ethics, nature of, 23-55
Military managers, origins of, 60
Military protest, types of, 185-199; appeal to higher orders, 193-94; relief, 191-93; refusal, 194-99; resignation, 188-91
Military service, vocation of, 19
MILPERCEN, 69, 97
Minimal application of force, 166
Missionary corps, 134, 216
Mollering, John, 145
Motto, Army, West Point, 71, 72, 157

Moral bearing, virtue of, 7
Moral: cowardice, 204; example, 224; obligation, 28; promise keeping, 29; reasoning, 20, 29, 123, 164, 180; responsibility, 186; sensitivity, 157; values, 28
Mutiny, 59-65
My Lai, 4, 40-41, 158, 191

Napoleon, 101
Naval Academy, 162
Nazi Germany, 158
NCO Academy, 212
Nelson, Lord Admiral, 163
Nixon, Richard, 92
Nontoleration clause, 31

Oath of commissioning, 71
Obedience, 161-62, 175
Objections to ethical codes, 42-50; codes vs. laws, 42-43; conflicting precepts, 44-45; content, 43; relative vs. absolute ethics, 46-48
Obligations, nature of, 29-30; as actions, 30; as expectations, 34; as group membership, 34; as social process, 33
Obligations, limits of, 178-183
Obsequium, 73, 158, 196
Occupationalism, 68
Officer Comrades Courts Of Honor, 214
Officer corps, 66
Organizational corruption, 71
"Ought implies can," 30, 125

Patriotism, 160-61
Patton, George, 86
Perfect virtue, 153
Personnel turbulence, 13, 59
Philosophical egoism, 106

Plausible denial, 192
POW. *See* Prisoner of War
Power, 172-73
"Premodern" organization, 63
Prescriptive ethics, 36, 133
Price of belonging, 134, 168
Prima facie duties, 43-44
Principle of universality, 48
Prisoner of War (POW), 212
Professional ethic, 8, 25, 82-83
Protective association, 147
Prudential judgment, 28
"Push-button battlefield," 226
PX scandals, 4
Psychological egoism, 18

Realism, 166
Refusal of orders, 194-99
Regimental officer board, 220
Rehnquist, Justice William, 93
Relative ethics, 46-47
Relief from orders, 191-93
Religious brotherhood, 146
Replacement stream, 59
Resignation, 98, 184, 188, 191
Respondeat superior, 40, 190, 192
Responsibility, virtue of, 165
Ridgeway, Matthew, 185
Role of the soldier, 81-83
Roman Army, 101, 111
Rommel, Erwin, 200
Rosser, Richard, 100
Rowney, George, 184
Ryan, John, 155

Sacrifice, 159-60
Scharnhorst reforms, 215, 220
Schlesinger, James, 176
Search and destroy, 184
Shills, Edward, 81
Situational ethics, 25, 50-53, 98

Smith, Adam, 15
Smithian economics, 18-21
Social Darwinism, 16-21
Social worth doctrine, 7
Socrates, 26, 33, 187
Soldier's oath, 156
Somme, battle of, 81, 226
Sorley, Lewis, 83, 86, 120
Soviet Army, 220
Soviet honor court system, 221
Special obligation, 167-68
Spencerian ethics, 18
Squadron Officer Study, 1978, 75, 175
State within a state, 90-93, 176
Stockdale, James, 56, 75
Summers, Harry, 201-2
Symbolism, 223

Tactical walk, 168
Taylor, Maxwell, 73, 75, 119, 120, 121, 122, 150
Teaching ethics, 206-13; code, 209; dialectics, 210; history, 208; reasoning, 208
Technical competence, 172
Thermopylae, 144
Toffler, Alvin, 135
Toner, James, 180
Traits of character, 152

Uniform Code of Military Justice, 94
United States Air Force Academy (USAF), 132. *See also* Academies
"Up-or-out" system, 59-63, 67, 72
Upton, Emery, 108

Value-free technician, 115
Verantwortungsethik, 47
Vietnam War, 3, 64, 65, 98
Virtue, nature of, 150
Vocationalism, 96, 167
Von Moltke, Graf, 80, 169

Wakin, Malham, 42, 124
War crimes, 4
Warrior ethos, 98
Watergate, 176
Waterloo, 81, 226
Weigley, Russell, 108
Westmoreland, William, 74
West Point, honor codes, 132, 198; failures of, 135
West Point, motto, 155. *See also* Academies
"Will to dissent," 162
"Wolf-at-the-door" syndrome, 13
Working ethics, 27, 48, 147

Yamamoto, Isoroko, 125
Yamashita, General, 41, 126, 192

"Zero defects," 12
"Zero defects mentality," 13
Zumwaldt, Elmo, 215

About the Author

Richard A. Gabriel is Professor of Politics at St. Anselm's College in Manchester, New Hampshire. He is a former Army intelligence officer and a major in the United States Army Reserve assigned to the Soviet Division of the Directorate of Foreign Intelligence in the Pentagon. His earlier books include *The New Red Legions* (Greenwood Press, 1980) a two-volume analysis of the Soviet military; *Crisis in Command,* the first major critique of American military performance in Vietnam; and *Managers and Gladiators*, an analysis of military bureaucracy. Dr. Gabriel is also a Military Affairs consultant to the staff of the House Armed Services Committee.

LIBRARY OF DAVIDSON